V E R I F I C

GEOTECHNICAL GROUTING

A Report from the ASCE Committee on Grouting of the
Geotechnical Engineering Division
and papers presented at the ASCE Convention
in San Diego, California,
October 23-27, 1995

Geotechnical Special Publication No. 57

Edited by Michael J. Byle and Roy H. Borden

Published by the
American Society of Civil Engineers
345 East 47th Street
New York, New York 10017-2398

ABSTRACT

This volume, Verification of Geotechnical Grouting, consists of a committee report from the ASCE Committee on Grouting and relevant juried technical papers. The committee report presents discussions on planning of a grouting verification program, grouting methods, and verification methods. The latter part of the committee report includes discussion on selecting verification methods and a set of charts indicating the applicability of various verification methods by grouting method, goal of grouting and subsurface conditions. The papers present various case histories involving verification of geotechnical grouting.

Library of Congress Cataloging-in-Publication Data

Verification of geotechnical grouting : a report from the ASCE Committee on Grouting of the Geotechnical Engineering Division and papers presented at the ASCE Convention in San Diego, California, October 23-27, 1995 / edited by Michael J. Byle and Roy H. Borden

p. cm.

Includes bibliographical references.

ISBN 0-7844-0132-2

1. Grouting-Congresses.I. Byle, Michael J. II. Borden, Roy H. III. American Society of Civil Engineers. Geotecnical Engineering Division. Committee on Grouting.

TA755.V49 1995 95-41892

624.1'51—dc20 CIP

Library of Congress Catalog Card No: 95-41892
ISBN 0-7844-0132-2
Manufactured in the United States of America.

GEOTECHNICAL SPECIAL PUBLICATIONS

1) TERZAGHI LECTURES
2) GEOTECHNICAL ASPECTS OF STIFF AND HARD CLAYS
3) LANDSLIDE DAMS: PROCESSES, RISK, AND MITIGATION
4) TIEBACKS FOR BULKHEADS
5) SETTLEMENT OF SHALLOW FOUNDATION ON COHESIONLESS SOILS: DESIGN AND PERFORMANCE
6) USE OF IN SITU TESTS IN GEOTECHNICAL ENGINEERING
7) TIMBER BULKHEADS
8) FOUNDATIONS FOR TRANSMISSION LINE TOWERS
9) FOUNDATIONS AND EXCAVATIONS IN DECOMPOSED ROCK OF THE PIEDMONT PROVINCE
10) ENGINEERING ASPECTS OF SOIL EROSION, DISPERSIVE CLAYS AND LOESS
11) DYNAMIC RESPONSE OF PILE FOUNDATIONS— EXPERIMENT, ANALYSIS AND OBSERVATION
12) SOIL IMPROVEMENT - A TEN YEAR UPDATE
13) GEOTECHNICAL PRACTICE FOR SOLID WASTE DISPOSAL '87
14) GEOTECHNICAL ASPECTS OF KARST TERRAINS
15) MEASURED PERFORMANCE SHALLOW FOUNDATIONS
16) SPECIAL TOPICS IN FOUNDATIONS
17) SOIL PROPERTIES EVALUATION FROM CENTRIFUGAL MODELS
18) GEOSYNTHETICS FOR SOIL IMPROVEMENT
19) MINE INDUCED SUBSIDENCE: EFFECTS ON ENGINEERED STRUCTURES
20) EARTHQUAKE ENGINEERING & SOIL DYNAMICS (II)
21) HYDRAULIC FILL STRUCTURES
22) FOUNDATION ENGINEERING
23) PREDICTED AND OBSERVED AXIAL BEHAVIOR OF PILES
24) RESILIENT MODULI OF SOILS: LABORATORY CONDITIONS
25) DESIGN AND PERFORMANCE OF EARTH RETAINING STRUCTURES
26) WASTE CONTAINMENT SYSTEMS: CONSTRUCTION, REGULATION, AND PERFORMANCE
27) GEOTECHNICAL ENGINEERING CONGRESS
28) DETECTION OF AND CONSTRUCTION AT THE SOIL/ROCK INTERFACE
29) RECENT ADVANCES IN INSTRUMENTATION, DATA ACQUISITION AND TESTING IN SOIL DYNAMICS
30) GROUTING, SOIL IMPROVEMENT AND GEOSYNTHETICS
31) STABILITY AND PERFORMANCE OF SLOPES AND EMBANKMENTS II (A 25-YEAR PERSPECTIVE)
32) EMBANKMENT DAMS-JAMES L. SHERARD CONTRIBUTIONS
33) EXCAVATION AND SUPPORT FOR THE URBAN INFRASTRUCTURE
34) PILES UNDER DYNAMIC LOADS
35) GEOTECHNICAL PRACTICE IN DAM REHABILITATION
36) FLY ASH FOR SOIL IMPROVEMENT
37) ADVANCES IN SITE CHARACTERIZATION: DATA ACQUISITION, DATA MANAGEMENT AND DATA INTERPRETATION
38) DESIGN AND PERFORMANCE OF DEEP FOUNDATIONS: PILES AND PIERS IN SOIL AND SOFT ROCK
39) UNSATURATED SOILS
40) VERTICAL AND HORIZONTAL DEFORMATIONS OF FOUNDATIONS AND EMBANKMENTS

41) PREDICTED AND MEASURED BEHAVIOR OF FIVE SPREAD FOOTINGS ON SAND
42) SERVICEABILITY OF EARTH RETAINING STRUCTURES
43) FRACTURE MECHANICS APPLIED TO GEOTECHNICAL ENGINEERING
44) GROUND FAILURES UNDER SEISMIC CONDITIONS
45) IN-SITU DEEP SOIL IMPROVEMENT
46) GEOENVIRONMENT 2000
47) GEO-ENVIRONMENTAL ISSUES FACING THE AMERICAS
48) SOIL SUCTION APPLICATIONS IN GEOTECHNICAL ENGINEERING
49) SOIL IMPROVEMENT FOR EARTHQUAKE HAZARD MITIGATION
50) FOUNDATION UPGRADING AND REPAIR FOR INFRASTRUCTURE IMPROVEMENT
51) PERFORMANCE OF DEEP FOUNDATIONS UNDER SEISMIC LOADING
52) LANDSLIDES UNDER STATIC AND DYNAMIC CONDITIONS - ANALYSIS, MONITORING, AND MITIGATION
53) LANDFILL CLOSURES — ENVIRONMENTAL PROTECTION AND LAND RECOVERY
54) EARTHQUAKE DESIGN AND PERFORMANCE OF SOLID WASTE LANDFILLS
55) EARTHQUAKE-INDUCED MOVEMENTS AND SEISMIC REMEDIATION OF EXISTING FOUNDATIONS AND ABUTMENTS
56) STATIC AND DYNAMIC PROPERTIES OF GRAVELLY SOILS
57) VERIFICATION OF GEOTECHNICAL GROUTING

PREFACE

The use of grouting in geotechnical engineering applications has expanded greatly in the United States in recent years. Technologies have improved to the point where grouting can be used successfully in a great many diverse applications. The increase in applications has led to an increased need for technology to verify that the desired end has been achieved. Testing and investigative tools have improved at an equally rapid pace as have the improvements in grouting technology. This publication is directed at identifying some of the technologies that have been successfully used in the verification of the performance of geotechnical grouting and offering some perspective on their use. This is not a comprehensive or exhaustive treatise. There are many more thorough discussions of various test methods, several of which are referenced in this publication. It is hoped that this publication can serve as a guide giving the reader a basis for understanding and selecting methods of verification.

This volume contains a Grouting Committee Report on the verification of geotechnical grouting and a collection of six peer reviewed technical papers. The technical papers were selected to provide examples of verification applications. The committee report is the result of the work of a number of individuals, without any of whom this publication would not have been the document that it is. This project was begun by the Grouting Committee in 1993, then under the leadership of Chairman Ed Graf. Roy H. Borden, now chairman of the Grouting Committee, served as co-editor to assist Michael J. Byle to complete this committee report. The resulting document represents the combined efforts of many people. The authors of various portions of the document are listed below:

Part A Planning an Effective Program	R. Karol, K. D. Weaver, M. J. Byle
Part B Grouting Methods	
Compaction Grouting	M. J. Byle, Gannett Fleming, Inc. E. D. Graf, Grouting Consultant
Permeation Grouting	W. J. Clarke, Geochemical Corp.—cement L. G. Schwarz, Northwestern Univ.—cement R. M. Berry, Rembco Engineering—chemical J. M. Malone, North Carolina State Univ.—chemical
Jet Grouting	L. F. Johnson, Heller and Johnson G. Burke, Hayward Baker Co. A. D. Walker, Nicholson Construction
Fracture Grouting	E. D. Graf, Grouting Consultant E. Drooff, Hayward Baker Co.
Soil Mixing	R. H. Borden, North Carolina State Univ.
Reviewers	R. Borden, D. Bruce, E. Graf, F. Gularte, K. Weaver, J. Welsh

Part C Verification Methods

Mechanical Methods	M. J. Byle, Gannett Fleming, Inc. E. R. Colle, Geoconstruction Consultant
Chemical Methods	W. J. Clarke, Geochemical Corporation
Geophysical Methods	J. J. Bowders, University of Texas at Austin M. J. Byle, Gannett Fleming, Inc. R. D. Woods, University of Michigan
Hydraulic Methods	S. D. Scherer, The Concrete Doctor, Inc. K. D. Weaver, Woodward-Clyde Consultants
Reviewers	P. Aberle, R. Borden, D. Bruce, M. Byle, E. Graf, K. Weaver, J. Welsh

Part D Method Selection

Detecting Grout	M. J. Byle, Gannett Fleming, Inc.
Permeability	S. D. Scherer, The Concrete Doctor, Inc. K. D. Weaver, Woodward-Clyde Consultants
Density	M. J. Byle, Gannett Fleming, Inc.
Strength and Modulus	M. J. Byle, Gannett Fleming, Inc. J. Benoit, Univ. of New Hampshire
Reviewers	P. Aberle, R. Borden, D. Bruce, E. Graf, S. Scherer, K. Weaver, J. Welsh

Part E Summary and Conclusions	M. J. Byle, Gannett Fleming, Inc.
Tools For Selection	ASCE Committee on Grouting

Each of the technical papers submitted for consideration was sent to two reviewers and required two positive reviews to be accepted for publication in this volume. In the case of a split review, the paper was sent for a third review. A total of 13 papers were reviewed, with 8 accepted for publication. Of those accepted, 6 papers were revised and submitted for inclusion in this volume. All papers are eligible for discussion in the ASCE Journal of Geotechnical Engineering. All papers are eligible for ASCE Awards. The individuals listed below served as referees during the peer review:

P. P. Aberle	R. H. Borden	E. D. Graf	J. L. Kauschinger	L. G. Schwarz
R. M. Berry	D. A. Bruce	F. B. Gularte	R. J. Krizek	C. Vipulanandan
A. Bodocsi	E. R. Colle	L. F. Johnson	S. D. Scherer	K. D. Weaver

The members of the ASCE Committee on Grouting: R. Borden (Chairman), P. Aberle, A. Bodocsi, D. Bruce, M. Byle, E. Colle, J. Gould, E. Graf, F. Gularte, G. Illig, L. Johnson, R. Karol, J. Kauschinger, R. Krizek, M. Leonard, A. Naudts, S. Scherer, L. Schwarz, R. Smith, C. Vipulanandan, A. Walker, K. Weaver, J. Welsh

Michael J. Byle
Roy H. Borden

CONTENTS

VERIFICATION OF GEOTECHNICAL GROUTING
Table of Contents

VERIFICATION OF GEOTECHNICAL GROUTING

ASCE Committee on Grouting

Edited by Michael Byle and Roy Borden

INTRODUCTION

The purpose of this document is to provide owners, design engineers, specifications writers, reviewing agencies, testing laboratories, construction engineers and managers , and contractors with a common basis for understanding and/or selecting the appropriate method(s) for verifying the effectiveness of grouting. It presents brief descriptions of various methods and of their applications for assessing whether or not grouting operations have adequately achieved the preplanned objectives. This document is intended to be a tool to enable the use of grouting with greater confidence. Grouting is broadly defined as the placement of a pumpable material which will subsequently set or gel in pre-existing natural or artificial openings (permeation grouting) or in openings created by the grouting process (displacement or replacement grouting). Grouting can be further defined by describing the placing method, and the grout material. In all cases, grouts flow by gravity or are pumped under pressure to fill openings in soil or rock or to displace soft or loose soils.

Much of the work performed in geotechnical engineering, and more specifically grouting, cannot be directly observed and measured. Observing and monitoring the completed geotechnical work were suggested by Karl Terzaghi in 1936 when he stated "the accuracy of computed results never exceeds that of a crude estimate, and the principle function of theory consists of teaching us what and how to observe in the field" (Cedergren 1967). This philosophy also applies to the design and construction of grouting projects. Because of unknowns in the subsurface, the design of a grout program is often based on assumptions regarding these unknowns. Like all geotechnical engineering projects, the successful design of a grouting program requires an adequate level of exploration and the input of an experienced design team. Verification is an essential part of the design and should be included in the specifications and contract documents. Verification of the effectiveness is best determined by careful observation of the finished project and thoughtful engineering judgement and analysis of physical data obtained while performing the work.

PART A PLANNING AN EFFECTIVE PROGRAM

1. Verification Program Development

The success or failure of a grouting job is related to human as well as technical problems. It is most important that, prior to the start of grouting, 1) the problem has been defined in detail to the satisfaction of everyone involved, 2) agreement has been reached upon the criteria which define success or failure, and 3) the data which may be necessary for "before" and "after" comparisons are gathered prior to the start of grouting.

Prior to the start of a grouting job it is essential to define the existing problem in sufficient detail so that conditions representing a satisfactory solution to the problem can also be specified. It is most important that the owner, design engineer, and the grouting contractor agree fully with the specific purpose of grouting, with the conditions which define success or failure, and with the tests, observations and/or data which can be used to verify the grouting results. It should be understood by all concerned parties that the grout formulations, injection rates, pumping pressures and grout takes for each stage of each hole be continuously monitored, recorded and evaluated, and that appropriate changes be made during the course of the work, or the work may prove to be a grout disposal project rather than a grouting program.

The material into which the grout is injected (referred to below as the "formation") is always opaque to human vision, so the flow of grout into or through the formation cannot be monitored visually. The final location of grout, in all of the methods discussed in this publication, can be controlled to some degree by the practices and procedures used. Viscosity, gel or set time, water-cement ratio, jet pressures, rates of injection and other parameters should be set in the design and adjusted during construction to control the rate and distance grout can travel. Since soil and rock are not uniform and variations in them are not always predictable, there always remains some risk that the grout may find its way to some unintended location. The success or failure of a grouting operation cannot be determined until some evidence of performance is obtained.

When the purpose of grouting is to add strength (or stability) to a formation, there is generally no way to judge success visually. Often, grouting may be done to increase an existing safety factor, which cannot be verified visually. When the grouting operation is completed, records of grout take, setting times, grout pipe location, injection depths and other construction data are a valuable tool and in some cases may be sufficient indicators of the proper placing of grout. The increase in some soil strength or indirectly related soil parameter can sometimes be measured at least qualitatively to confirm a change in the soil strength.

Often, it is assumed that if the grout sets in the desired location, the job is successful and the success of the grouting operation is measured by field procedures which verify the grout location. The actual location of solidified grout in a formation can be verified or inferred by various field tests. Methods that measure the change in electrical conductivity or acoustical response have been researched with positive results. However, these methods require specialized equipment and expert personnel. Depending on the site conditions and application, some of these methods can be economical.

Unknown or unanticipated underground conditions may adversely affect the grouting performance. Detailed geotechnical investigation is the best method to avoid unnecessary surprises. However, it is unreasonable to assume that any level of investigation will identify every possible situation. When considering a monitoring program, one should consider the possible conditions that could affect the grouting. Ground water may dilute the grout, particularly if grout or ground water flow rates lead to turbulence. Fast flowing ground water could move the grout away from its intended location. Dissolved chemicals in the ground water may affect the grout. Discontinuities in the soil or rock mass could direct the grout to an unplanned location. Soil or rock strata, not noted in the soil borings, could affect the grout flow and behavior in unexpected ways. Careful monitoring and evaluation of grouting parameters during the course of the work can enable previously unknown conditions to be identified and treated appropriately.

When the purpose of grouting is to shut off or divert existing flow or seepage, the success of a grouting operation can sometimes be observed visually by observing the change in water flow. When the purpose is to prevent anticipated flow or seepage, success may be inferred at some future date if seepage does not occur. A simple arrangement of wells or piezometers can identify changes in ground water flows and gradients even while the grouting is in progress. For temporary cases, it is necessary for the owner, design engineer, and contractor to agree beforehand upon the amount of seepage reduction that constitutes success or failure.

2. Goals of Verification

In planning an effective program it is important to clearly define the properties to be measured. Verification may include direct measures of performance of the grouted system or may include methods for verifying proper construction. The verification program goals should be set appropriately for the project. In most cases only a qualitative assessment of grouting performance is needed. But, in some cases, accurate quantitative assessment of the grouting may be needed. The cost of any method of verification should be balanced against the benefit to be gained.

Poorly defined verification goals can lead to confusion. If verification is provided as an afterthought without a clear goal, disputes can arise concerning whether the grouting has been successful. It is important when selecting verification methods that the right questions are asked:

What is the purpose of the grouting?
What measurable changes are expected due to grouting?
What methods can measure these changes?
How will the verification test results be evaluated?
What will be the acceptance criteria?
What are the consequences of failure?
What is the cost for the verification?

Other questions may arise as to the timeliness of results, how the verification fits into the construction sequence, and other issues.

Well defined verification goals will arise from the answers to all of these questions. A good verification program will be cost effective, provide an appropriate level of assurance, provide results in a reasonable time frame, and should use the simplest most reliable technology possible.

3. Verification by Design

The time to begin the verification program is in the planning stages of the project. Depending on the application, the selection of the grouting method may depend on how well the results of that method can be verified. Verification methods should be selected consistent with project goals and incorporate the verification in the design and constructability evaluation. Specifications should be established for the grouting, monitoring and verification testing to provide the required level of quality assurance. It is important to consider interruptions to the grouting that may be needed and to allow for before and after testing if needed to measure improvement. Test grouting should include the same verification measures intended for the production grouting. Excavated test pits may be used to calibrate and verify the results of indirect tests in the test grouting stage. The project schedule must include sufficient time for the completion and evaluation of the test grouting prior to production grouting and for evaluation of the verification test results. It is of much less value to get the verification report indicating unacceptable areas after the grouting contractor has demobilized from the site than to get real time data..

The verification equipment and procedures must be made integral to the grouting program and other construction activities on the site. Where before and after testing is necessary, consider the need to protect measuring equipment left in place. Include redundant equipment in high hazard areas. Consider the construction site environment and its effect on the proposed equipment and tests where necessary. If a test section is desired, include it in the specification and include the verification testing for the test section.

PART B GROUTING METHODS

1. Introduction

In its broadest meaning, the term “grouting” is used to describe the process of injecting a material into a geologic formation. The reasons for grouting are many and significantly influence the methods and materials used in a particular application. Soil grouting techniques are usually described as being in one of the following categories (Figure 1.1):

(1). Compaction grouting
(2). Permeation grouting
(3). Jet grouting
(4). Fracture grouting
(5). Soil mixing

Grouting techniques have been applied in support of open-cut excavation and tunneling activities around the world. They have also been widely used in settlement remediation, underpinning, groundwater control, embankment stabilization and providing adequate bearing capacity for structures. The following paragraphs in this chapter provide a brief description of each method.

2. Compaction Grouting

2.1 Definition

In 1980, the ASCE Grouting Committee, in its glossary of terms, defined compaction grout as: "Grout injected with less than 1 in. (25 mm) slump. Normally a soil-cement with sufficient silt sizes to provide plasticity together with sufficient sand sizes to develop internal friction. The grout generally does not enter soil pores but remains in a homogeneous mass that gives controlled displacement to compact loose soils, gives controlled displacement for lifting of structures or both." Over the years, those involved with compaction grouting have realized that slump alone is not an adequate measure of injected grout behavior.

Grouts with slumps greater than 1" have been used successfully on many occasions and grouts with excessive clay fines have unintentionally fractured the ground forming lenses even though the slump was on the order of 1". Recently, the term "mobility" has been used to describe the tendency of a grout to flow under pressure (Warner, 1992). Regardless of slump, the formation of grout bulbs as opposed to lenses is enhanced by the use of low mobility grouts.

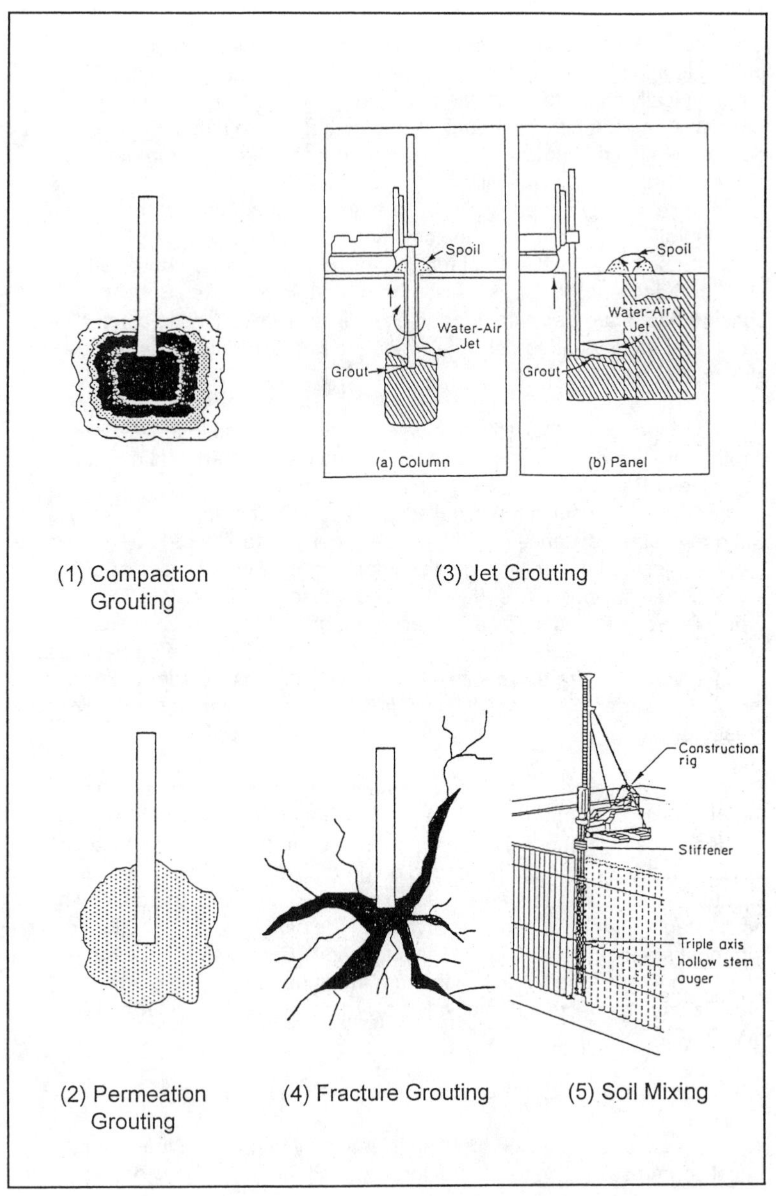

Figure 1.1 Soil Grouting Techniques

2.2 Method

Compaction grouting involves the use of high pressure positive displacement pumps to inject a very stiff mortar-like grout into the subsurface soils. The grout typically consists of silty sand or silty gravelly sand, cement and water. Bentonite or other clay commonly is added to facilitate pumping. Care must be exercised in adding clay, as increased amounts tend to promote potentially undesirable mobility in the ground Other additives may be used to enhance strength or control set. The grout is generally mixed on the site with pugmill or auger mixers. In some cases, where very large quantities of grout are being used, the grout may be mixed at an off-site concrete batch plant and delivered to the site in standard ready-mix trucks. Special considerations are necessary for off-site mixing applications to control the grout consistency and to delay set when the rate of grout take is reduced by site conditions.

Injection pipes are either drilled or driven into the ground to the desired depth. Inside pipe diameter is chosen to match equipment and project objectives. Pipes 38 mm (1.5 in) or larger in inside diameter have been used successfully. The pipes are installed to fit tightly in the holes. The pipe is withdrawn some distance (stage), on the order of 1 to 2 m (3 to 6 ft), and grout is injected by a positive displacement pump. Grout is injected until a specified refusal criterion is reached. The refusal criteria may include pressure, volume and surface displacement or combinations. Typical refusal criteria may be 3.5 MPa (500 psi) pressure at the collar of the grout pipe at a given low pumping rate, specified heave (typically on the order of 3mm) of the ground surface, or a specified volume of grout injected at some minimum pressure.

Successive grout stages may be either accomplished from the bottom up or the top down. The bottom up, or up-stage, procedure is the most common and least costly method. The up-stage method requires that the pipe be installed to the bottom of the zone to be grouted and successive stages are injected in a continuous manner as the pipe is withdrawn. The top down, or down-stage, method is done by advancing the pipe only to the bottom of the top-most stage and grouting the top stage first. Successive stages are grouted one at a time by drilling through the upper completed stage(s). The grout is usually injected at relatively slow rates, generally less than 0.06 m^3/min (2.0 ft^3/min).

2.3 Design Considerations

The purpose of compaction grouting is generally to compact loose soils by radial compression and to lift structures and attendant services on the surface. This may be done for a variety of soil types and conditions. Compaction grouting is generally most effective in sandy or silty soils, but has been effectively used in trash and rubble fills, loose unsaturated sandy clays and other c-ϕ soils. Compaction grout has also been used to compact fills composed of clay "clods" and displace soft clays, close off subsurface erosion paths and to density collapsed materials in Karst formations. Compaction grouting is generally not recommended in saturated soils of low permeability where excess pore pressures may develop, thus preventing proper compaction. However, structures have been releveled by displacing soft clays that were undergoing consolidation. Compaction grout has also been used to densify peat pockets in soft bay muds to reduce differential settlement.

The selection of the grouting method will depend on both cost and performance requirements. The down-stage method has the advantage of providing very good control of the grouting process. This is important when grouting in close proximity to structures or where it is important to verify the grout location. The price for this added control in down-stage grouting is additional drilling costs and a longer construction schedule. Up-stage grouting may be more cost effective and provides adequate control for most applications.

Refusal conditions should be established with consideration for the anticipated goal of the grouting. High volumes of grout injected at single locations may be necessary when filling voids, but may cause damage to subsurface structures or uncontrolled ground heave in soil if not properly monitored. The volume of grout injected at each stage is a function of the injection spacing, soil conditions, the rate and pressure of injection and other site constraints. The degree of control during compaction grouting is inversely proportional to both the rate of injection and the slump/plasticity of the grout. The grout program should include split spaced secondary holes between primaries to verify the effectiveness of the grouting.

Grout strength is not a major factor in most applications. The grout need only have a strength equal to that of the surrounding densified soil. Typically, a nominal strength of a few MPa (hundred psi) is all that is needed. For special applications where grout columns are constructed, higher strengths similar to structural concrete are generally used.

2.4 Properties

Grout properties are similar to those of weak concrete. Compaction grouts containing cement have unconfined compressive strengths in the range of 3-28 MPa (400 to 4000 psi) depending on the mix design. The grout injection will density or displace the surrounding materials. The modified materials may have a higher density. In the case where grout displaces soft soils in Karst, the surrounding rock may behave more monolithically.

2.5 Construction Monitoring

Monitoring during grouting is essential to verify proper performance of compaction grouting. As a minimum, the grout consistency, injection rate, injection pressure and injected volume in each stage should be monitored and recorded. This data should be reviewed to evaluate the ground and grouting performance. Injection pressures will be lower and injected volumes higher in softer or less dense areas. Secondary injections should show lower volumes and higher pressures because of the improvement induced by primary injections. Additional split spaced holes may be used to verify the success of secondary grouting.

The drilling or driving of grout injection pipes should also be used as a tool to evaluate the grouting. Drilling or driving will encounter greater resistance in improved soils or when encountering grout. Drill cuttings will also give additional evidence of the presence of grout.

3. Permeation Grouting

3.1 Definition

Permeation grouting occurs when grout fills soil pores or rock fissures without causing significant movement or fracturing of the soil or rock formation. It refers to the replacement of water and/or air in the voids between the soil particles or rock surfaces with a grout fluid at pressures chosen so as to prevent fracturing in the grouted soil mass. Permeation grouting has been used as a means for soil stabilization in many engineering projects, where cohesionless soils required an increase in strength or reduction in permeability to be utilized for the intended purpose. It is applicable in coarse to fine sands, depending on the grout mix and soil properties

3.2 Methods

Permeation grouting typically requires mixing of the grout components, pumping of the grout to a manifold or injection pipe and injection through an open ended pipe or sleeve port pipe. Some single component grouts do not require mixing at the site. Where mixing is required it is typically done using a continuously metered process, but can also be done as a batch process.

Grout may be injected through drill rods, hollow stem augers, or a pipe driven or drilled into the ground. Where driven pipes are used, a knock-out disposable tip is provided to prevent soil from plugging the pipe.

Sleeve pipes are pipes that have ports at intervals with rubber sleeves covering the holes. The rubber sleeves prevent the intrusion of soil into the pipe but expand under pressure to permit grout flow. Sleeve pipes are usually placed in bored holes and are grouted into the holes with a weak grout. Inflatable packers are used to isolate an individual injection port during grouting, and/or water testing. The sleeve pipes can be washed out after grouting and reused for multi-level injections or multiple injections at a single port.

The grout injection process is controlled by the set or gel time of the grout, the rate and pressure of injection. Accurate meters are necessary to measure the volume of grout injected and may be needed for controlling the mix proportions and gel time of some grouts. Pressure gages should be of sufficient accuracy and precision to measure the full range of pressures expected. Pumps vary by type of grout to be injected, and may be "moyno", piston or diaphragm type.

The permeation grout is a fluid of sufficiently low viscosity that it can permeate the porosity of a soil or rock matrix. The fluid grout must be injected at relatively low pressures to prevent hydraulic fracturing of the matrix and undesirable heave of the ground surface.

3.3 Design Considerations

It is conventional for preliminary estimates of groutability to be based on soil permeability or void size. Table 3.3.1 and Table 3.3.2 are examples of grout selection tables developed by AFTES (1991).

One rule of thumb suggests that soils with less than 1% to 2% fines (minus 200 sieve material) are easily groutable, soils with 2% to 20% fines may be only moderately groutable, and soils with 20% to 25% fines are only marginally groutable (Baker, 1982). The major factors controlling groutability are the pore size and pore size distribution of the soil and the viscosity of the fluid grout. Less viscous lower strength grouts would be able to permeate formations with smaller pore sizes than would more viscous higher strength grouts. For soils with fines contents in excess of 5 to 15%, depending on intended grout viscosity, it may be useful to perform bench scale tests or preliminary field injections to evaluate groutability.

Table 3.3.1. Types of Grouts Used, Depending on the Type and Size of Fissures (after AFTES, 1991)

Type and Size of Voids	Types of Grouts
Open Voids; karsts; wide fissures	Cement-based grouts with coarse fillers (gravels) Cement-based cellular grouts Quick setting grouts
Large fissures (average opening >1cm)	Cement grouts with fine fillers (ash, fine sand filler) Quick setting grouts Cement-based cellular grouts Bentonite, clay, cement grouts Polyurethane foam Carbamide
Average fissures (1mm to 1cm)	Pure cement or bentonite Clay-added grouts Synthetic foams Resins
Fine fissures (0.5 to 1mm)	Special improved penetrability grout Silicate gel Acrylic resins
Very small fissures (less than 0.5 mm); porous material	Deflocculated bentonite Low-viscosity silicate gel Acrylic resins Phenol resins

Based on the material components, permeation grouts can be divided into two categories: chemical grouts and cementitious grouts.

3.3.1 Chemical Grouting

Chemical grouts are primarily used for stabilizing granular materials and for treating finely fissured rock or concrete. The objective may be to provide increased strength or to retard water seepage. For sandy soils, chemical grout may be used to convert a cohesionless soil into a cohesive or a nearly impervious material. Grouting with chemicals involves the filling of the voids in soil or rock with a fluid chemical which sets or cures to alter the properties of the geologic mass. The rate of chemical injection and set time are such that the soil particles are not significantly displaced by the injection process.

Table 3.3.2. Fields of Application of Grouts for Granular Soils (after AFTES, 1991)

Legend: Normal field of application; Limited by cost

GROUT			Strengthening (C) or Watertightening (w)
CEMENT			C
CLAY-CEMENT			WC
GROUT with filler Cellular Grout			WC
CLAY GEL BENTONITE (deflocculated, strengthened)			W
GROUTS with improved penetration			WC
EMULSIFIED BITUMEN			W
SILICATE GEL	Stren-gthening	concentrated	C
		low viscosity	C
	Watertigh-tening	concentrated	W
		very diluted	W
RESINS	ACRYLIC		W
	PHENOLIC		C

GROUND PROPERTIES							
Initial Permeability k in (m/s)	10^{-7}	10^{-6}	10^{-5}	10^{-4}	10^{-3}	10^{-2}	10^{-1}
	Coarse pre-treated alluvial. Fine alluvial (gravels and sand, sands, silty sands)				Coarse grounds scree. Coarse alluvial.		

Typically, property enhancement is an increase in shear or compressive strength accompanied by a decrease in permeability.

Chemical grouts are generally composed of three primary components: (1) matrix forming "base materials", (2) "reactants" or "reagents", and (3) "accelerators" or "retarders". The relative concentrations of the three components can vary significantly depending on the application and desired properties. Karol (1982) suggested that the most widely used chemical grouts could be divided into the following chemical families: Sodium silicate formulations, Acrylamides, Lignsulfites, Phenoplasts, and Aminoplasts. Additional grout types include: Ureas, Polyurethanes, and Epoxies.

In terms of grout volumes employed in geotechnical structural chemical grouting, Baker (1982) estimated that sodium silicate based chemical grouts had been at that time used in over 90% of the cases.

Selection of the particular chemical grout is of major importance to the performance of the system. Significant items for consideration include:

- nature of the application; i.e. size of voids or fractures and type of substrate
- potential handling problems including flammability
- chemical resistance, reactivity and permanence of the grout due to environmental factors
- toxicity or other environmental issues
- viscosity and flow characteristics of the grout
- dilution effects from groundwater
- set time and curing characteristics of the grout system
- syneresis characteristics
- cost

One of the advantages which may accrue from chemical grouts is the broad range of properties potentially available:

- strengths vary from low strength soft gels or foams to high strength
- viscosities at injection time vary from near water to thick oil
- stability to acids, bases or organics can vary from zero to infinite
- flexibility of the grout can vary from soft sponge to rock brittle
- adhesive properties vary from very low to extremely high
- water or moisture effects on cure and other physical properties are broad

3.3.2 Cement Grouting

Cementitious grouts are composed of cement and water, and often contain additives such as clay, Bentonite, sodium silicate, dispersants, retarders and accelerators, as appropriate. Most cement grouting operations are completed with ordinary Type I Portland cement. Other cements used for grouting include blast furnace slag and finely ground cements.

Standard Portland cements are satisfactory for grouting 300 micron fissures and fine pebbles. The finely ground cements may be used to grout 30 micron fissures and medium sands. The viscosity of cement grout can be modified with a variety of additives available on the market today. Bleeding of cement grout can be controlled with additives such as Bentonite.

Effectiveness of grouting with a particulate grout depends mainly on the following factors:

- size of the grout particle with respect to the soil pore diameter or fissure width
- grout viscosity
- grout stability
- injection pressure
- pumping rate.

The main consideration for slurry grouting to seal cracks and fissures in rock, or to be injected into soils for either water control or structural improvement purposes is the grain size of the particulate grout compared to the width of the rock fissure or pore size of the soil to be grouted. Accordingly, reference is frequently made to a variety of groutability ratios (N). The following ratios for soils and rock have been suggested (Task Force 27, 1990):

For soils:
N = (D15) soil / (D85) grout
N > 24 grouting consistently possible
N < 11 grouting not possible

Nc = (D10) soil / (D95) grout
Nc > 11 grouting consistently possible
Nc < 6 grouting not possible

Where, for example, (D15) soil is the grain size for which 15% of the soil by mass is smaller and (D85) grout represents the grain size for which 85% of the grout particles are similarly smaller.

For rock:
Nr = width of fissure / (D95) grout
Nr > 5 grouting consistently possible
Nr < 2 grouting not possible

The above ratios are indicative of guidelines in the literature. However, in practice, it can not be assumed that groutability ratios larger than the "not possible" values will ensure successful grouting. Successful grouting depends on the careful selection of the grout which is most compatible with the geologic mass being treated together with injection techniques.

4. Jet Grouting

4.1 Definition

Jet grouting is a means of hydraulic cutting and mixing in situ soil materials with a fluid grout injected at high velocity to create a stabilized mixture of soil and grout.

4.2 Method

Jet grouting creates a stabilized mixture of soil and grout by cutting the in-situ soil materials with a fluid mixture, which is injected at a high velocity and with a pressure on the order of 35 to 40 MPa (approximately 5,000 to 6,000 psi). Two specialty components are used to perform this work: a drill manufactured specifically for jet grouting that provides access to the in situ soil and consistency during hydraulic cutting and mixing; and a pumping unit to deliver the fluid(s) at the appropriate volumes and pressures.

The technique uses a drilling system to gain access to the soils requiring modification and is most often designed as a series of interconnected columns of various lengths and geometry to create a mass of mixed soil and grout. The method is applicable to a wide range of soil types and has been applied to simple stabilization, underpinning, excavation support, anchors, hazardous waste containment, groundwater control, slope stabilization, erosion protection, and various shaft and tunneling projects.

There are basically three types of jet grouting systems available in the U.S. Each system injects a high pressure liquid to cut or mix the soil in place. After the drill has advanced to the proper depth, injection of the components begins and the slow lift and rotation of the drill rods creates a cemented zone of soil. The single and double rod systems use a cement slurry pumped at high velocity to both cut the soil and form the soil grout mixture. These systems rely on mixing the disturbed soil with the cement slurry. The double rod system includes an air component to aid in increasing the cutting distance by conically shielding the high pressure slurry to achieve additional energy at distance from the rods. The triple rod system uses the combination of high velocity water shielded in a cone of air to cut and displace the soil to the surface while simultaneously injecting the disturbed area with a cement slurry from a lower nozzle. The volume of soil ejected is normally greatest with the triple jet system because the excavation operation is separated from the cementing operation. Also, the air shroud improves the cutting efficiency of the water jet, and the exiting of the air from the hole tends to improve soil removal efficiency by lifting the soil.

The typical jetting parameters vary with the three methods. The single fluid system parameters are the water/cement ratio of the grout, grout volume and pressure, number and sizes of nozzles, the drill rod rotation rate and lift speed. Parameters for the double fluid system include air pressure and air flow rate in addition to those for the single fluid system. Parameters for the triple fluid system include those for the double system plus water volume and pressure and the number and sizes of water jets. Additionally, in all systems the grout may contain additives, such as, Bentonite, fluidizers, air and fly-ash.

4.3 Design Considerations

The strength of the soil and grout mixture is vital in underpinning and tunneling applications. Specified compressive strength should be based on analysis, since the specifying of an unnecessarily high compressive strength will greatly increase costs. In most underpinning applications, an unconfined compressive strength of 2 to 3.5 MPa (300 to 500 psi) is more than adequate. When soil treatment is combined with conventional shallow foundation design, high strengths are not needed and it may be more appropriate to use strengths consistent with bearing pressure requirements. Relatively small amounts of organic soils will significantly reduce strength. To achieve high strengths in organic soils, the jet grouting must flush out most of the organics.

The dimensions and location of jet grout columns are most important in forming ground water cutoffs and in underpinning. In all cases field tests should be performed to verify the contractor's selection of jetting parameters. If site conditions allow, the contractor will excavate and measure the test column. The jetting parameters may then be varied, if necessary to achieve required size or shape.

The permeability of the soil and grout mixture is most important in cutoff and containment applications. As with specifying strength, the specifying of permeability should be realistic or the project will carry unnecessary expense or be impractical. Also, the primary source of seepage may occur at construction details, such as the penetration of tieback anchors through cutoff walls or windows between adjacent columns if overlap is insufficient.

4.4 Properties

The properties of jet grouted columns are a function of many factors, including the system used, the jetting parameters, water table location, curing time, and most importantly, soil characteristics. Depending on the application, relevant properties to specify for jet grouting include dimensions, location, strength and permeability. Design compressive strengths in the range of 2 to 10 MPa (300 to 1500 psi) are easily obtainable in most inorganic soils. Permeabilities on the order of 10^{-6} cm/sec are generally achievable and in some conditions permeabilities as low as 10^{-8} cm/sec are possible.

4.5 Construction Monitoring

In some cases, it is not possible to excavate test columns due to their depth, the presence of the water table or other site constraints. In these cases sufficient information can usually be obtained by installing feeler probes or wells at varying distances from the test columns, or, following grouting, by drilling and either obtaining core samples or recording drilling resistance. In addition to these techniques, automatic data-logging equipment can be used to monitor injection pressures, grout volume and, in the case of triple-rod systems, the volume of effluent removed from the hole.

In underpinning applications it is important to check and prevent the tops of the jet grout columns from settling. An adequate hydraulic head of grout maintained through or adjacent to the foundation will prevent this from occurring. Bleeding of the grout can be reduced by the use of additives. Bentonite is typically used for this purpose, however, a reduction in column strength occurs. Finally, jet grout columns are sequenced to avoid fluidifying too much soil beneath a building at a given time.

5. Soil Fracture Grouting

5.1 Definition

Fracture grouting is the intentional fracturing of soils using grout pressures high enough to fracture the soil at the point of injection. The basic concept is to use a grout material that will not permeate the soil. Although chemical grouts are occasionally, but rarely, used, most frequently used are Portland cement grouts. Fracture grouting is sometimes used in rock grouting and this is addressed.

5.2 Method

In fracture grouting, stable fluid grout is injected under high pressure with the intention of introducing controlled fracturing of the ground. Repeated injections after periods of curing tend to densify the adjacent ground, decrease the local permeability , stiffen and strengthen the soil due to the hardened grout lenses, and provide the capability to lift footings or maintain existing elevations during an adjacent excavation.

Because the process requires that the soil be fractured and not necessarily permeated, soil fracture grouting may be used in most soil types ranging from weak rocks to clays. However, because of a rather broad range of compaction and permeation techniques for coarse grain soils, soil fracture grouting has found a particular niche in treating cohesive soils.

The injection of grout to produce lenses has been performed both with open-ended pipes and with sleeve-port pipes. In the latter case, a system of injection pipes is introduced into soil in such a way that openings spaced at distances ranging from 300 mm to 1000 mm along the length of the pipes, are uniformly distributed in the soil layer. The boreholes containing the grout pipes are commonly arranged in fan-like arrays or in parallel rows. A packer capable of creating a tight seal within the grout pipe, thus isolating one zone for injection, is pushed to the selected sleeve port. Injection is generally started with predetermined injection batches, the volume of which depend on the volume of soil to be stabilized. To achieve uniform stabilization, injection continues in those zones in which the recorded injection pressures did not reach the required level during the first or second injection stage. In loose soils, fracturing will not occur until the initial injections densify and stiffen the zone. Prevention of soil heave has to be ensured by monitoring at regular intervals

5.3 Design Considerations

Soil fracture grouting is used to increase the shear strength and resultant bearing capacity resistance of soils as well as to raise structures. To date, the largest application for the process has been to remediate against settlement caused by soft ground tunneling under structures. Grout pipes are positioned based on the bearing pressure distribution resulting from the structures to be supported and on local soil stratification. A system should provide sufficient redundancy to allow for pipe blockages and grout injection rates should be sufficient to achieve heave at a rate greater than the maximum predicted rate of settlement of a structure.

Ground improvement by soil fracture is based on three mechanisms.

- The soil unit or skeleton is reinforced by a series of hard grout lenses which propagate out from the injection point to form a matrix of hard grout and soil.
- The fluid grout finds and fills voids and causes some compaction in

more coarse grained soils along the grout lenses created.

- The plasticity index of saturated clays decreases through the exchange of calcium ions originating from cement or other fillers.

Unlike compaction grouting, soil fracture grouting, when performed with relatively small volumes of localized injections, will induce only minimal increases in pore pressures in cohesive soils and as such will produce little or no long-term consolidation. However, should a mechanism for consolidation persist, the grout placement pipes are entirely reusable enabling additional grouting to re-level a structure or further reinforce the soil without the installation of additional pipes.

Fracture tightening of rock is accomplished by the use of grout pressure that will move the rock within the formation by enlarging the fissures being grouted and thereby reducing the size of the fissures not being grouted. This method has the goal of strengthening the formation structure and/or reducing the permeability of the formation. The technique is particularly valuable when using an unstable grout (high bleed) because, when the grout pressure is released, the rock pressure will squeeze the bleed water into the fissures too fine for the grout.

5.4 Properties

Cement based grout mixes are most commonly used for fracture grouting and must be designed to minimize excess bleed water and shrinkage. If using chemical additives, grout gel times must be readily controllable to a tolerance of approximately 10 seconds in order to control grout placement in discrete areas after repeated injections at short intervals. Grouts will often contain fillers, sand or other admixtures, depending on the application. When sleeve-port grouting techniques are used, consideration must be given to the more restrictive nature of pumping grout through small openings.

The constituent proportions of a cementitious grout mix will vary greatly depending on the application. For instance, for the compensation of settlement due to tunneling, the grout need only be as strong as the existing soil. However in weak soils where an improved bearing capacity is required, added grout strength is more useful. Flowability and set time of grout are also very important but can have a wide range based on the specific requirements of a project. Due to the heterogeneous nature of the injected soil mass, it is necessary to use methods that investigate a large soil volume to verify improvement.

5.5 Construction Monitoring

To control the fracture process, it is very important to continuously monitor and evaluate injection pressure, rate and total injected volume. In addition to grouting information, an integral part of the process is the ability to monitor movements of affected structures. A movement monitoring system consistent with the engineer's criteria for settlement and angular distortion of the structure must be implemented. Where long arrays of grout pipes are used, pipes should be accurately surveyed for alignment to verify injection point locations.

6. Soil Mixing

6.1 Definition

Soil mixing is a process by which cementitious materials are mechanically mixed with in-situ soil using a hollow stem auger and paddle arrangement. In concept, the process attempts to do by mixing what high pressure jets accomplish during jet grouting.

6.2 Method

Available configurations include single shaft augers (0.5 to 4 m in diameter) or gangs (2 to 5 shafts) of augers about 1m in diameter. The single-row multiple shaft auger is generally used for in situ soil mixed walls while double-row multiple shaft augers are used for areal treatment of soft or contaminated ground.

As the mixing augers are advanced into the soil, grout is pumped through the hollow stem of the auger shaft and injected into the soil at the tip. The auger flights and mixing paddles blend the soil with the grout. When the design depth is reached, the augers are withdrawn with simultaneous rotation, allowing the mixing blades on the rotating shafts to further mix the soil and grout to form continuous stabilized lime, soil-cement, or soil-cement-Bentonite columns or panels. Depths of improvement are limited only by the available equipment. The Swedish lime columns are typically 10-15m deep, while depths greater than 30 m have been achieved in the United States and more than 60 m in Japan. By successive soil mixing, a wide variety of desired shapes of treated soil can be produced.

Kawasaki et al. (1981) present a table showing the performances and work execution conditions of a deep mixing machine (Table 6.1).

Table 6.1. Performance and Work Execution Condition of Deep Mixing Machine (after Kawasaki et al., 1981)

<table>
<tr><td colspan="2">Depth of Stabilization</td><td>30 - 40 (m)</td></tr>
<tr><td colspan="2">Improved Area</td><td>4.26 - 5.74 (m^2)</td></tr>
<tr><td colspan="2">Penetration Velocity</td><td>1.0 -2.0 (m/min)</td></tr>
<tr><td colspan="2">Withdrawal Velocity</td><td>1.0 - 1.5 (m/min)</td></tr>
<tr><td rowspan="2">Rotating Speeds of Blades</td><td>during penetration</td><td>20 - 30 (rpm)</td></tr>
<tr><td>during withdrawal</td><td>40 - 60 (rpm)</td></tr>
<tr><td colspan="2">Cement Content (for 1 m^3 of soil)</td><td>1.4 - 3.0 (kN/m^3)</td></tr>
</table>

6.3 Design Considerations

In situ soil mixing techniques are applicable to a range of soft alluvial and marine soils, organic soils and reclamation fill. Typically, clay soils with SPT "N" values of about 4 blows or less and granular soils with "N" values of about 10 blows or less can be treated. With heavier equipment and electric or hydraulic motors with higher torque to rotate the shafts, soils with "N" values of about 10 to 30 blows respectively can be treated (Toth, 1993).

Soil mixing has been used to address a wide variety of problems, including settlement control, slope stability, soil heave prevention, earth retention and liquefaction risk reduction.

6.4 Properties

The hardening agent used in soil mixing could be cement or lime, slurry or powder. The most commonly used dosage of cement slurry lies between 100 and 200 kg per cubic meter of native soil and can reach, in some cases, 300 kg/m^3. The strength of soil cement is dependent on the type of soil treated, mix design of the grout, and degree of mixing. Unconfined compressive strengths between 100 and 5000 kN/m^2 have been reported (Kawasaki, et al, 1981). The deformation modulus of stabilized soil will also increase greatly. The final strength of improved non-cohesive soil will be reached after 28 days, while in cohesive soil, strength gains have been reported up to 90 days after mixing.

6.5 Construction Monitoring

As the augers and mixing tools prescribe the zone of improvement, the most critical factors to insure successful applications of soil mixing include quality control of the grout components, rate of injection of the grout to verify the desired dosage, and monitoring of rates of auger insertion and withdrawal, as well as rotation rates for mixing tools. When barriers are constructed for seepage control, alignment of augers during insertion also must be carefully monitored to avoid "windows" between columns or panels.

PART C VERIFICATION METHODS

1. Introduction

A wide variety of testing methods can be used to either directly or indirectly measure the performance of grouting. These include mechanical, chemical, geophysical, hydraulic and other methods. The focus of these methods is to determine a change in some property of the subsurface after grouting or to detect the presence of grout. Since we cannot see into the ground to know where the grout has gone or that it has achieved the desired goal, we must use other methods to supplement the construction monitoring. Some of these methods are non-intrusive and/or non-destructive and can be used without disturbing or damaging the grouted area. The non-destructive methods generally are indirect and require interpretation of the desired information from some other measured property. Some methods require computer analysis to reduce the data. The more intrusive methods can either collect samples for viewing or other analysis, or measure some in situ properties. Some methods can be used while grouting is underway and others are sensitive to disturbance cause by ongoing construction and can be used only when all operations have stopped. All of these factors are important when designing a Quality Assurance Program for grouting.

2. Mechanical Methods

Mechanical methods are probably the easiest to understand, since intuitively, it is easy to imagine a device that either obtains a sample or takes some direct measurement of or in the ground These methods are typically used for geotechnical investigations or construction monitoring for common civil engineering projects. These include strength tests, movement monitors, and density tests.

2.1 Load Testing

Load testing is a common approach to measuring the performance of foundation systems Load tests can be conducted on plates which model foundation loads or on structural elements such as compaction grouted, jet grouted or soil mixed columns. The approach is simply to load a plate or foundation element using dead weight or a hydraulic jack. The hydraulic jack reacts against either a dead weight or anchors of some sort. The anchors may be piles, tie downs or other foundation elements that can resist the test load. Load tests can also be performed in tension where tensile resistance is an important factor (Bruce, et al., 1995).

2.2 Penetration Resistance

The consistency of a soil is commonly evaluated by pushing or pounding a probe into the soil and measuring the force needed to advance the probe. There are two basic approaches: static and dynamic. The static methods advance a probe by applying a static force to the probe to advance it. The dynamic methods apply an impact of known energy and count the number of impacts needed to advance the probe.

2.2.1 Static Methods

The most common static method is the cone penetrometer test (CPT). The CPT is often termed "quasi-static" since the force is measured while the probe (cone) is being advanced at 10 to 20 mm/s (2 to 4 ft/min). There are many variations of the CPT but most differ in the method of measuring the applied force and differentiating between the side friction and point bearing of the probe. The two main types of CPT devices are the mechanical cones and the electrical cones. Both the mechanical and the electric cones typically have conical points with a 60° point angle and a projected areas of 10 cm^2 (1.55 in^2). Modern CPT friction cones have a friction sleeve which is the same diameter as the conical tip. The force needed to slide this sleeve through the hole created by the tip is measured. The mechanical cones use push rods to first push a tip section and then the friction sleeve into the ground. Electric cones use strain gages and/or load cells to measure the force in lieu of the hydraulic pressure acting on the rods. the electric cone can measure friction and tip resistance simultaneously. Other more advanced electric cones may include inclinometers and piezometers for additional data on alignment and pore pressures. Some new cones include options such as geophones for seismic testing, electrodes for resistivity testing, or infrared equipment for identifying organic compounds and ground penetrating values (Bratton, et al., 1995)

The CPT data consist of force measurements for tip resistance versus depth. Friction cones provide additional friction resistance data. The results of the cone resistance, friction resistance and the ratio between friction and cone resistance are plotted side by side at the same scale versus depth. These data can be interpreted to derive soil parameters. By comparing the test results before and after grouting, the improvement can be measured.

The CPT has the advantages of being relatively quick, low cost and that it provides nearly continuous data over the full depth of the probe. The disadvantages are that it cannot penetrate very hard or dense materials such as rubble, large gravel and some hardened grouts and it is a test that only evaluates a very small plan area.

2.2.2 Dynamic Penetration Test Methods

Dynamic methods are among the oldest and simplest methods in use. Dynamic penetration tests using cones or other types of probes enjoy wide use, but the Standard Penetration Test (SPT) is probably the single most widely used method of evaluating soil. The SPT test (ASTM D 1586) uses a 63.5 kg (140 lb) hammer falling through 0.76 m (30 in) to drive a 35 mm I.D. split barrel sampler. The number of hammer blows is recorded to advance/drive the sampler 0.15 to 0.3 m (6 to 12 in). The resistance values are indicative of the soil stiffness or relative density and are locally correlated with a variety of soil properties. The method is typically used in boreholes. A disturbed sample of the soil is obtained and can be visually examined or submitted to laboratory testing.

Many other dynamic penetration tests are available including smaller hand held penetrometers, some of which provide samples and some that do not. Correlations for these other penetrometers are not so widely available and may need to be developed for each application. For some large projects it may be feasible to consider developing correlations for use of lighter penetrometers where only shallow testing is needed.

Penetrometers are generally point tests, that is, they test only the material in the immediate vicinity of the point of the probe or sampler. Boring methods can influence the results of penetration tests. Methods such as jetting, wash borings, or other methods can disturb the soils to a significant depth below the sampling point. This can cause spurious penetration values. Baker (1982) indicates that SPT does give some indication of strength increase in chemically grouted soils, although results are usually highly variable.

2.3 Probing and Sampling

Probing and sampling remain the most cost effective procedures for locating a grouted formation mass. Sampling procedures are of course performed after the completion of grouting. In contrast, probe tests must be done both before and after grouting, because comparative data is needed.

Probes may be any solid bar which will withstand driving forces. (Although static penetration tests are possible, dynamic tests are simpler to perform and interpret). The driving force (the weight used and the distance it is dropped) must be consistent throughout the "before" and "after" tests, so that comparisons will be valid. Generally, the compared data will be the number of blows to penetrate a specific distance.

Probe tests give good to excellent results in fine-grained soils. They obviously cannot be used in fractured or solid rock, or in soil formations containing rock particles large enough to deflect or halt the probe. Also, test results become increasingly difficult to interpret with increasing depth, as the lateral resistance of the soil increases and becomes the major factor in probe resistance.

Automatic monitoring of drilling parameters has been used to successfully interpret subsurface conditions. This involves using recorders or computers to record down pressure and torque during drilling. These data are correlated with geotechnical data and used as a means to obtain subsurface information during drilling of grout holes.

When probes are ineffective, soil and rock sampling procedures may give positive data. Shelby tubes are effective in loose fine-grained soils solidified with a weak grout. However, some grouted soils can be disturbed excessively by the insertion of a Shelby tube. Dense soils , strong grouts and rock inclusions will generally preclude the use of thin walled tubing for sample recovery. Heavy walled tubing can often be driven through formations not amenable to Shelby tubes. However, the samples recovered will usually be fractured to the point where it is difficult to verify that the formation has been grouted. The same is true for drilled samples. In addition, the liquids sometimes used in drilling may hide or disguise the presence of grout particles.

Sampling of chemically grouted sands, generally , has been found to be most successful by taking carved samples from test pits. Chemically grouted soils are generally disturbed using most push type samplers (Shelby Tube, Pitcher Sampler) because of the brittle and low strengths usually achieved. One successful primitive method is monitoring the rate of advance of a hand-held wet-head drill to detect the presence of grouted sands.

When sampling by any method, the identification of grout particles in the loose state can be assumed to mean that the sampled portion of the formation has been grouted. Chemical grout particles are lighter in weight than sand or rock particles, and will settle on the top of a liquid column. Sedimentation tests are useful in identifying grout in field samples, and in the drill fluid.

2.4 Excavation/Coring

Excavation and coring are methods used to extract and observe subsurface conditions. Excavation of test pits is commonly used to observe in-place grouted columns installed by jet grouting, chemical or compaction grouting, to expose the treated subsoil for observation or testing, or to obtain samples for laboratory testing. Test pits are only suitable for relatively shallow depths and are typically used for depths less than 7 m (20 ft). Deeper test pits maybe used where proper shoring is used. The cost of test pits increases substantially when shoring is needed. Test pits can be dug using hand tools or mechanical excavating equipment depending on the depth and nature of the soils to be tested. Block samples may be carved from within the excavation with minimal disturbance.

Coring is the process of cutting an undisturbed sample from the subsurface by drilling with a hollow bit. The sampler cuts away the annulus around the specimen recovered. This may be done using a hollow auger with a special cutting shoe in soils or a rotary core barrel in rock or other consolidated materials. Coring will tend to disturb soft soils and weakly cemented materials such as chemically grouted soil. Coring in consolidated materials generally requires the use of a drilling fluid to lubricate the cutting shoe. Air, water and drilling mud are commonly used drilling fluids. The effect of these fluids on the core specimen should be considered in using this method. Shallow cores up to about 1m (3 ft) may be obtained with relatively light equipment. Deeper coring generally requires a heavy duty drilling rig. Soil coring with a Pitcher sampler, Denison sampler, and electric diamond core have been used to obtain samples of chemically grouted soil (Davidson and Perez, 1982).

The operation of coring into the into grouted soil can create microfracturing of the sample. The microfracturing caused by coring into grouted soil can greatly increase the values of permeability obtained by laboratory testing.

For jet grouting applications, sampling devices are available which allow retrieval of wet samples from any depth immediately after column construction. Also, casting 75 mm (3 inch) diameter schedule 80 PVC into the wet soil and grout mixture and retrieving it the next day is an effective method of sampling in some soils.

2.5 Modulus Tests

The Flat Dilatometer and Pressuremeter are devices that directly evaluate the modulus of soils or rock. Both generally require the use of a drilling rig, but the flat dilatometer may be also be used from a cone penetrometer rig. The flat Dilatometer Test (DMT) consists of a flat blade with a circular diaphragm on one side. After inserting the blade into the soil, the diaphragm is loaded internally by pneumatic pressure, and the pressure necessary to move the diaphragm against the soil (termed the lift-off pressure) is measured. The movement of the diaphragm is monitored by a contact switch within the blade. Because the DMT is normally hydraulically advanced into the soil, its use is limited to soils which are free of gravel or other obstructions.

The pressuremeter is similar to a large packer in its operation. The conventional pressuremeter generally consists of three inflatable cells; two guard cells and one measuring cell. The measuring cell is inflated with water under pressure. The quantity of water forced into the cell is used as a measure of the volume, and hence the displacement of the cell. The pressure on the water is a measure of the resistance of the surrounding soil or rock. The guard cells, placed above and below the measuring cell, are expanded with air and act as confinement to minimize end effects. The middle cell deforms as a cylinder, uniformly loading the sides of a borehole. The most common pressuremeters are used in predrilled holes to calculate the stiffness of any material in which a freestanding hole can be maintained. In some circumstances, drilling mud is needed to maintain an open hole. These pressuremeters can be used in most soils and soft rocks. However, sharp inclusions can puncture the pressuremeter membrane making it difficult for use in gravel or rubble.

In some cases where submerged loose sand or other conditions cannot permit a stable open borehole, self boring pressuremeters can be used to obtain geotechnical parameters. Self boring pressuremeters can be used in most soils and soft rocks and are advances either by cutting, using various cutters and bits, or by jetting (Benoit et al.,1995)

Also, laboratory testing of undisturbed core samples of grouted soils has been used to measure changes in soil strength and modules. Numerous authors have reported the unconfined compression tests and triaxial shear tests to measure strength and permeability of chemically grouted sand.

2.6 Extensometers

Extensometers are instruments that measure longitudinal deformation. They may be used to measure the settlement of a layer of soil or to measure deformation of a structure or the ground surface. There are many type of extensometers. The simplest extensometers may consist of two stakes or pins between which the distance is measured periodically. The changes in this distance reflect movements. An extensive discussion of various extensometers is presented by Dunnicliff (1988).

The accuracy of extensometers varies widely among the various types. An arrangement of two stakes driven into the ground on opposite sides of a crack is a simple extensometer which is monitored by periodic measurement of the distance between the two stakes. The accuracy of such a system is subject to the precision of its construction, that is how rigidly the stakes are anchored, how accurately the distance is measured, how precisely the index marks are placed, and the sensitivity of the equipment to the environment. On construction sites, sites subject to vandalism or heavy traffic areas, the stakes could be easily damaged, requiring re-installation and loss of continuity in the monitoring.

The next level of extensometer instruments includes crack gages, which may be of the pin and caliper type or plastic grid and overlay type. These are reasonably reliable and inexpensive to purchase, install and monitor. Crack gages are typically used to measure movement across an existing discontinuity which may be a construction joint, saw cut or crack usually in concrete, masonry or rock.

Larger extensometers include tape and wire extensometers for measuring changes over larger distances. Some extensometers use ultrasonic waves to measure distance between two points or Linear Variable Differential Transformers (LVTD's) or Direct Current Differential Transformers (DCDT's) for measuring relative movement.

2.7 Tiltmeters

Tiltmeters measure angular movement or rotation of surfaces. A simple tilt measurement is to measure horizontal deviation over the length of a plumb line or spirit level. More sophisticated devices such as the portable clinometer and electronic tiltmeters using accelerometers or vibrating wire technology are also available. The accelerometer based electronic instruments can be permanently mounted and read remotely. Long term degradation may be a problem for some vibrating wire equipment.

2.8 Settlement Plates

For measuring vertical movements of the ground surface, optical surveying of surface mounted survey points is a widely used and relatively inexpensive method. The survey points are typically constructed as metal plates anchored in the ground with a vertical rod extending up to the ground surface. Often a 0.5 m square steel plate is used with a standard pipe fitting welded to it. An advantage of this method is that the vertical rod can be protected by an outer pipe sleeve and extended upward as fill is placed, thereby separating the settlement of the subgrade from compression of the fill. This may be desirable for applications where it is necessary to monitor the performance of improved subgrade under newly placed fill. The accuracy of this method is limited by the accuracy of the survey, fixity of the reference benchmark and integrity of any extensions added to the plates. An in depth discussion of these methods for deformation monitoring is presented by Dunnicliff (1988).

Remotely monitored liquid settlement transducers (LST's) are available for locations where vertical access to the monitoring point is not feasible. These are appropriate for use under structures or in areas where settlement plate rods would be damaged or too obtrusive for the site use. LST's consist of a steel plate attached to a sealed container which is connected by tubing to a remotely located fluid reservoir. The relative head between the reservoir and the sealed container is measured with a pneumatic or electronic pressure transducer. The reservoir elevation must either be fixed or regularly surveyed and the plate elevation is determined by subtracting the differential head from the reservoir elevation. Some problems can occur due to air bubbles forming in the tubing or instrument, however, this can be corrected by de-airing the fluid system. These LST's are also subject to all of the limitations of pore pressure transducers. The system must also be charged with an antifreeze solution to prevent freezing in cold weather applications.

2.9 Fluid Levels

For active short term monitoring of level changes, a fluid level can be used. Fluid levels consist of a reservoir and one to several measurement tubes. The system operates on the principal that a fluid will seek the same level in the reservoir and in the tube. With a scale attached, they can be used to measure relative elevation between the tube location and the reservoir. These are commonly used in compaction grouting and slab jacking to monitor lift and contour floors. Fluid levels are generally quite accurate but can be affected by air bubbles in the lines, and constriction of the tubing. Care should be taken when using fluid levels for long term applications, since evaporation, fluid degradation, freezing and other factors can affect their accuracy.

2.10 Rotating Laser Levels

Rotating Laser Levels consist of a laser that impinges upon a rotating prism which causes the laser beam to scribe a horizontal plane. The laser is detected by instruments which indicate deviation from the plane of the laser. This method is typically used during construction to monitor vertical movements of structures, slabs or the ground surface.

2.11 Density tests

Density can be readily measured using standardized testing procedures including the nuclear density gage (ASTM D 2922), sand cone test (ASTM D 1556), and less commonly, rubber-balloon method (ASTM D 2167), drive cylinder method (ASTM D 2937) and the drive sleeve method (ASTM D 4564). All of these tests except the nuclear method operate by extracting a known volume of soil and determining its mass and thus directly determine the in situ density.

The nuclear method operates on the basis that a more dense material will deflect or scatter more gamma radiation than will a less dense material. The relationship is nonlinear and the instrument must be calibrated. The method is indirect and no theoretical approach predicts the relationship between gamma ray transmission and the system components. Calibrations are generally provided by the manufacturers of the nuclear gages. However, the calibration may not hold true for all soil types, particularly micaceous soils.

2.12 Laboratory Shear Strength Tests

Laboratory strength tests provide a direct strength measurement. Laboratory tests require obtaining undisturbed samples of grouted soils from the site. Laboratory strength tests include unconfined compression, triaxial compression and direct shear tests. Where it is not feasible to obtain undisturbed samples, reconstituted samples can be prepared using the same grout mix and soil samples (ASTM D4320). However, such test results may not accurately reflect field conditions due to the non-uniformity of in situ grout penetration and the variations in natural soil structure.

The triaxial compression test consists of applying an axial compressive load to a sample which is enclosed in a membrane and subjected to a confining pressure. Standardized test methods are available for triaxial testing of soil (ASTM D2850) and rock core (ASTM D2664). The shear strength parameters, "c" (cohesion) and "ϕ" (angle of internal friction) are interpreted from three or more triaxial tests by plotting the Mohr-Coulomb failure envelope. The unconfined strength can be directly evaluated from unconfined compression tests for soil (ASTM D2166), rock cores (ASTM D2938) and chemically grouted soil (ASTM D4219).

Direct shear testing for soil (ASTM D30801) and rock (USACE 1970) consists of shearing a soil or rock sample along a predetermined plan. The specimen is subjected to a confining stress normal to the direction of shear. Shear strength may be directly measured for a single confining stress.

Similar to triaxial compression tests, c and ϕ are obtained from the Mohr-Culomb plot for three or more tests.

3. Chemical Methods

3.1 pH Indicators

Many grout materials are either more alkaline or acidic than the soil materials or ground water. pH indicators can be used to identify the presence of grout where an indicator is selected that changes color at a specific pH indicative of the grout. For sodium silicate grouts with a high pH, phenolphthalein is commonly used. Phenolphthalein changes from colorless to pink at a pH between 8.4 to 9.6. In typical applications such as in grouted sand, Phenolphthalein would be sprayed onto a sample or the soil exposed in an excavation and the pink areas would indicate the presence of grout components.

3.2 Chemical Dyes

Chemical or food dyes can be used to identify flow of water through a grouted zone. The dye is typically introduced to the water at some point up-stream of the zone of interest. Some dyes are very intense colors while others are colorless but glow under ultraviolet light. Dyes may also be added to some chemical grouts to make them more visible for verification purposes. Dyes can be added to grouts to evaluate travel and dilution in water shutoff applications to aid in the selection of an optimum gel time. Wilson (1968) presents detailed procedures for dye tracing.

4. Geophysical Methods

Geophysical test methods are those tests which measure physical properties of the ground. Commonly, these properties include stress wave propagation velocity, electrical resistance or conductivity, gravitational properties, radar reflectivity, etc. These methods typically measure properties which can indirectly be related to the action or presence of grout and can thus be used for verification of grouting. The most common geophysical methods are Seismic, Electrical, Electromagnetic, Acoustic Emissions and Micro-gravimetric methods. Each of these is discussed in more detail below.

The high cost and destructive nature of physical-mechanical test methods coupled with the increased desire for subsurface data are leading us to increasingly rely on geophysical techniques. There are many advantages to geophysical methods including:

- Rapid information,
- Non-destructive,
- No cuttings to dispose of,
- Continuous data vertically and laterally, and
- Multiple types of information can be obtained with one technique

Technological advances have eliminated many of the disadvantages, but there are a number of limitations remaining. The most notable of these are:

- No physical samples are obtained
- These methods have limited experience or history in grouting applications
- Methods are specific to site conditions
- Can be costly for some applications
- Interpretation is subjective for some methods
- The results are not always conclusive

A listing of current and potential applications for the most common geophysical methods is presented in Table 4.0.

4.1 Seismic Methods

Methods that measure the propagation of stress waves, commonly called seismic waves, are included under the general heading of seismic methods. Seismic methods measure the speed (velocity), frequency and amplitude variations of induced seismic waves. The waves are induced either as a pulse, using a hammer blow, or explosive charge, or as steady-state vibration using a wave generator. The ground motion is commonly measured with transducers, called geophones or accelerometers. The ground motions are recorded with a seismograph, oscilloscope, or other signal processor. The seismic waves are small amplitude vibrations induced into the ground and can be overwhelmed or obscured by other vibrations. Therefore, seismic methods may be sensitive to background vibrations caused by construction equipment, compressors, traffic and other sources. As a result, seismic testing is frequently performed when there is no construction activity, such as at night or on weekends. Seismic methods are not practical for locations where excessive background vibrations cannot be eliminated during the test unless sophisticated wave from stacking equipment is used.

Seismic waves are divided into two basic types: body waves and surface waves. There are two types of body waves: dilational waves (compression waves or p-waves), and distortional waves (shear waves or s-waves). There are several types of surface waves, but only the Rayleigh Wave is widely applied. Each of these wave forms will be present in a surface generated

Table 4.0 - Geophysical Methods and Their Applications

Geophysical Method	Current Applications	Potential Applications	References
Seismic	Voids, Stratification discontinuities, Youngs modulus, shear modulus, and other elastic constants.	Grout strength, grout location and extent.	Ackman & Cohen (1994) Byle, et al. (1991) Davidson and Perez (1982) Greenhouse, et al. (1995) Woods and Partos (1981)
Electrical			
Conductivity Resistivity	Liquids, voids, density detection & mapping	Grout location	Ackman & Cohen (1994) Komine (1992) Greenhouse, et al. (1995)
Electromagnetic			
Ground Probing Radar	Voids, Liquids, Changes in dielectric constant	Grout location and continuity	Koerner et al. (1981) Ackman & Cohen (1994) Bowders et al.(1982)
Microwaves	Near-surface voids, Seepage	Voids near borehole	Koerner et al. (1978) Koerner et al., 1982.
Very Low Frequency	Voids or density anomalies at depth	Voids at depth	Ackman & Cohen (1994)
Acoustic Emissions	Density changes, Flow of Liquid, Mass movement		Koerner, et al. (1984) Koerner, et al. (1985) Huck, et al. (1982)
Gravimetric	Voids or density anomalies at depth	Voids	Greenhouse, et al. (1995)
Magnetic	Voids or high density inclusions	Voids	Greenhouse, et al. (1995)

pulse. Rayleigh waves are important for evaluating surface effects such as the influence of vibrations due to impacts, explosions or construction on adjacent buildings but can also be used to determine stratification that may be created by grouting. Compression waves generally travel the fastest and are easiest to detect as first arrival signals in unsaturated soils. The rates of travel of p- and s- waves are directly related to the modulus of the medium through which they travel. Where grouting can be expected to alter the modulus of the ground, p-wave or s-wave studies may be most appropriate. P-waves can travel through water, even pore water, but shear waves can not, because water has no shear strength. Therefore, shear waves may be necessary to evaluate grouting below the groundwater surface where the p-wave velocity of the grouted soil is less than that of water.

Seismic waves travel radially from the source, but may be reflected or refracted by discontinuities or variations in the ground. It is important in the interpretation of any seismic method to know the expected path of the waves being measured. The waves generally travel faster in stiffer media. It is possible for a wave to travel from the surface down to a stiffer stratum, along the stiffer stratum and then back up to the surface to arrive before a wave travelling directly along the ground surface. In fact, this is the basis for refraction surveys.

In all seismic methods the key parameter measured is the wave velocity. Velocity is measured by timing the travel of a wave from the source to a geophone or between two geophones along paths of known distances. The soil moduli are related to the seismic wave velocities as follows:

$$V_s = \sqrt{G/\rho} \rightarrow E = G(3\lambda + 2G)/(\lambda + 2G)$$

$$V_c = \sqrt{(\lambda + 2G)/\rho} \rightarrow G = V_s^2 \rho$$

Where:

V_c = Compression wave (p-wave) velocity

V_s = Shear wave (s-wave) velocity

λ = Lame's constant

G = Shear modulus

ρ = Mass density of medium

E = Young's modulus

Advantages

Seismic methods can measure average improvement in soil parameters over wide areas. This can be useful for ground improvement methods that produce non-uniform densification or stiffening of the subsoils. Seismic methods require the seismic wave to travel some distance through the soil or rock and hence can be varied to examine almost any size area under the right conditions.

Limitations

The accuracy of seismic methods is limited by the following:

- Accuracy of the test equipment
- Accuracy of distance measurement between the sensing locations
- Variations in the subsurface conditions
- Validity of the ray path assumptions

Modern seismographs and geophones are generally quite accurate and not usually a problem. Over short distances, the measurements between the source and the geophone must be very precise. One millimeter in 10 meters is not as significant as 1 millimeter in 1 meter. The direction and orientation of layering, buried structures, or inclusions in the soil will alter the ray path from that anticipated. In thinly stratified soils bedded along the path of the test, only the stiffest layers will be measured.

Seismic methods are only useful for verification where the grouting will result in a measurable increase in the overall stiffness of the soil or rock mass, or in the stiffness of zone to be tested. These methods cannot be used where underground structures are within the zone of influence and could affect the shortest ray path. Highly irregular rock surfaces, such as in karst, can sometimes cause distortions that can complicate interpretation of wave arrivals.

4.1.1 Direct Transmission (Borehole Methods)

The simplest methods to understand and interpret are the direct transmission methods. Direct transmission involves the measurement of seismic waves travelling on a direct path between the wave source and the geophones. When using direct transmission methods it is important to remember that the wave path can deviate from a straight line, especially at long distances. The accuracy of the method is directly related to the accuracy of measuring the distance between source and geophone. For direct transmission applications below the ground surface, the test must be done from boreholes. There are three different configurations for borehole seismic testing: up-hole, down-hole and cross-hole. Schematics of the three arrangements are shown in Figure 4.1.1. The up-hole and down-hole methods have the advantage of only requiring one borehole, whereas the cross-hole method requires two. The up-hole method has been found to yield poor results due to interference from the sides of the borehole where the signal source is a hammer blow on a rod (Byle et al., 1991). A directional hammer blow may be used to generate shear waves at the ground surface or in the subsurface with a directional downhole hammer. Reversing the

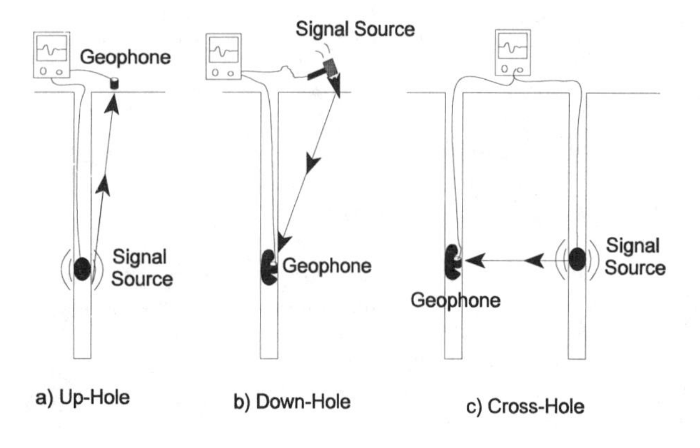

Figure 4.1.1 Direct Transmission Seismic Methods

direction of the hammer blows will invert the shear waves but not the p-waves. This makes for easier identification of the shear wave arrival. For short spacings between the source and geophone and deep boreholes, the borehole alignment should be verified with an inclinometer. For crosshole tests, borehole alignment must be verified. Crosshole tests should be peformed as specified by ASTM D4428/D4428 M-91.

4.1.2. Refraction

Seismic refraction is a common procedure that operates from the ground surface and can be used to measure changes in the subsurface profile and material moduli. The method requires that the following assumptions be met:

- The seismic wave velocity increases with depth
- Contacts between layers are generally planar

These assumptions can generally be met for residual weathered deposits and most other natural deposits where the degree of consolidation and soil stiffness increases with depth. Refraction is not appropriate for soft zones confined beneath denser soils or for highly variable surfaces such as karst.

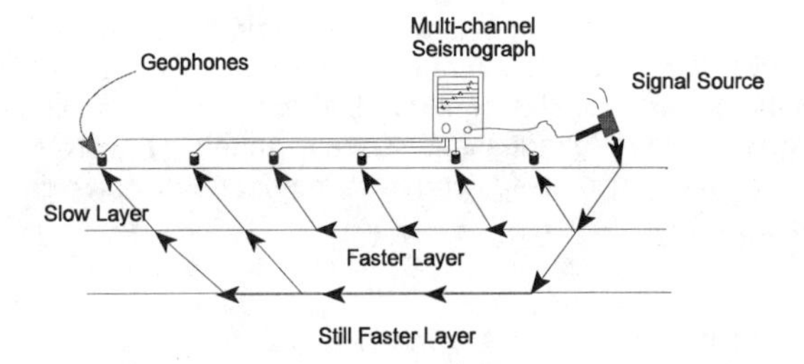

Figure 4.1.2 Seismic Refraction

Refraction is based on the principle that the first arrival of a seismic wave will follow the quickest path to the geophone from the source. As the spacing between the source and the geophone is increased, the fastest path will be a direct path through the surface layer until a longer path through a stiffer (and hence faster) zone at depth results in a shorter travel time. If the surface zone is stiffer, the fastest path will always be through the shallower layer. The general arrangement for this test is shown in Figure 4.1.2.

4.2 Electrical Methods

Geophysical methods that measure electrical properties of the subsurface materials include resistivity, conductivity and EM methods The application of electrical methods for verification of grouting is very limited, and is based upon the difference in the electrical resistance of water and fluid grout in the interstices of a porous medium. Verfel (1989) has pointed out that the small penetration of the electrical measurement as compared to the distance of the grout penetration presents a disadvantage in the use of this method. Komine (1992) has demonstrated that electrical resistivity of chemically grouted sands is dependent on the proportion of the voids filed by cured grout and has proposed use of resistivity tomography for field measurements. Though this method holds promise, it is not in widespread use.

4.3 Ground Penetrating Radar

Ground penetrating radar (GPR) is similar to the radar used by law enforcement and that used in aircraft, but uses longer wavelengths to penetrate the ground. GPR is generally limited to shallow depths and cannot see through clays well. Gradual and subtle changes in the ground cannot be detected. The method is useful for evaluating near surface voids or inclusions.

In a GPR system, a transmitter antenna emits a short pulse of electromagnetic energy (10 to 1000 MHz) directed into the ground. The pulse propagates through the ground and is reflected, refracted, diffracted or scattered and some of the energy is detected by the receiver antenna. The transmitter and receiver may be located in the same antenna (typical for surface applications) or may be separate (typical in borehole applications where the transmitter is in one borehole and the receiver is in another). The time for the electromagnetic pulse to travel through the ground and reflections of the pulse off materials with differing dielectric constants are recorded and used to image the subsurface in terms of boundaries with contrasting electrical properties. Dielectric properties and conductivity of the subsurface materials control the propagation of the radar pulse (Okrasinski et al., 1979).

High resolution can be achieved with GPR because of the high frequency of energy used (Bowders et al., 1982). However, depth of penetration may be severely limited when water-saturated zones, clay zones, or highly conductive material such as salt or man-made metallic objects are encountered. These zones either attenuate or reflect most of the energy. Depending on conditions, the depth of imaging may be less than a meter to as great as 300 m (Barbin, 1993).

For grout verification applications, GPR in the downhole or crosshole technique is most appropriate. Before and after grouting travel times are measured and the relative effectiveness of the grout is assessed. Conceptually, the velocity of the radar pulse increases as subsurface voids are filled with grout. Thus, radar travel times are decreased as the mass becomes monolithic.

In air, the radar pulse travels essentially at the speed of light, 3×10^8 m/s. The velocity in any other medium is determined by the ratio of the speed of light and the dielectric constant of that material as shown below (Eq 1).

$$v = \frac{c}{\sqrt{\xi_r}} \quad (1)$$

where v = velocity in medium of interest, e.g., soil or rock

c = speed of light, 3 x 108 m/s, and
ξ_r = dielectric constant of the medium of interest,
e.g., water ~ 80, dry sand ~ 2.

While GPR has not been extensively used for verification of grouting, it does have a history of use. In 1980, crosshole radar was used to measure the

change in porosity of a fractured rock mass before and after grouting at the Cross Power Generating Station in South Carolina.

In 1982, pre- and post-grouting crosshole GPR surveys for sections of the Pittsburgh light rail tunnel were performed(Parish et al., 1983). In this instance, the change in travel times between pre- and post-grouting was used to assess grout location.

Limited application of GPR to grout verification has taken place since the early 1980's partially attributable to the need, at that time, for highly experienced personnel for system operation and more importantly for data interpretation (Baker, 1982). Since the early 1980's there has been significant advancement of GPR equipment including signal enhancement (Kurtz, 1995) and data reduction and interpretation features (Olhoeft, 1993). Also there has been widespread application of GPR during the intervening period (Hauser et al., 1995; Tsuchida et al., 1995).

Advantages of the GPR technique are that it can produce high-resolution imaging of the subsurface under optimal conditions and that it operates in real time. However, GPR equipment is expensive and subsurface conditions may not be conducive to imaging or may limit the depth of penetration of the energy. The latter limitation is overcome by closely spacing the boreholes used for downhole survey and grout injection.

4.4 Acoustic Emissions

Acoustic emissions (AE) are generated by stress waves within materials during dynamic processes. The particular dynamic process may be the result of an externally applied stress, such as that produced by grout pressure resulting in hydrofracturing or, it may be the result of some other unstable situation, such as seepage of grout through cracks and voids. Some examples of acoustic emissions are the cracking of wood when it is overstressed, the "crying" of tin as it is bent, the cracking of ice and the crunching of snow. A stress (force) is applied, something gives (a strain is produced), and energy is released that appears partly as sound energy.

The detection and monitoring systems for acoustic emissions in metals

(Lord, 1975), rock (Blake et al., 1974) and soil (US EPA, 1979) are well advanced. The instrumentation system consists of a wave guide (potentially a grout pipe), a pickup transducer, an amplifier, and a readout system. The readout system is often a frequency counter but can also be based on amplitude or RMS for continuous emission monitoring. The frequency response of the transducer is of major importance and can vary over wide limits. In metals, transducers are typically responsive in the 1,000 to 1,000,000 Hz range; in soils, accelerometers are used with a frequency response from 500 to 8,000 Hz; in rock geophones with a frequency response from 1/4 to 100 Hz are used.

The most suitable frequency range for monitoring seepage flow (grout movement) depends greatly on the manner of pickup. If the pickup is made directly at the source of the emission, that is downhole or in the grout itself, the pickup transducer can be of high frequency which will eliminate background noise. Conversely, the farther away from the source of the emissions, the lower the frequencies which must be utilized. This permits for longer transmission of the emissions along the wave guide without attenuation of the signal.

Koerner et al., (1978) provide a case history in which AE was used to locate seepage zones beneath a small (3.6 m high) earth dam. The AE results were correlated with the findings from a drilling program. The seepage paths correlated well with the AE activity. In another case history, AE was used to monitor the travel path and location of slurry grout (Koerner, 1984, 1985). In this case, multiple pickup transducers were used to simultaneously record the acoustic emissions generated by grout flowing though the subsurface. By triangulating the arrival times of the emissions, the investigators were able to determine the direction of the grout movement from the grout pipe.

Acoustic emissions can also be used to detect when hydrofracturing is occurring in a formation that is being grouted. This technique can then be applied to allow site specific grout pressure to be used rather than rules of thumb.

4.5 Tomography

Tomography is an analytical technique for evaluating geophysical data such as the speed and amplitude of sound waves traveling through the earth, or electrical resistivity, radar frequency and signal strength, etc. between two boreholes. The principle of tomography is similar to that used by a CAT-scan, or a magnetic resonance image (MRI). Images can be created by numerical analysis of the data for a multitude of ray paths and using super-position. The method handles large volumes of data and requires that the transmitters and receivers completely surround the study area as much as possible. This is accomplished by placing instrumentation in boreholes and along the ground surface to send and receive the energy pulses, seismic waves, electromagnetic waves, or electrical current. This requires specialized equipment and large computational capability. It has been found possible, by use of this method, to detect grout plumes 75 mm and greater in thickness, and to trace grout travel along selected pathways. Seismic tomography has been found effective in detecting grout plume changes in before-and-after surveys (Yen and Gutierrez, 1993).

5. Hydraulic Methods

5.1 Introduction

Hydraulic methods include equipment and procedures related to measuring the behavior of water beneath the ground surface. Of interest are the ground water elevations, hydraulic gradients and the permeability of the ground.

Elevations and gradients are usually obtained by using open pipes or other devices which measure the water pressure at specific locations. Gradients are evaluated by measuring the change in head between points. This may be measured in both the horizontal and vertical directions.

The permeability of soil (here used as a general term for unconsolidated deposits, irrespective of origin) is one of the most difficult properties in geotechnical engineering to measure. Permeability most commonly is expressed in cm/sec, although other units of velocity such as feet per day or feet per year also remain in use. The range of permeabilities of natural deposits may be as much as 10 orders of magnitude, from 1×10^{2} in clean, open gravel to 1×10^{-8} cm/sec or less in clay. For the purposes of this report, the discussion of permeability will focus on groutable soils and rock masses, i.e., on materials with a permeability of 1×10^{-5} cm/sec or greater.

This section describes some of the equipment and procedures used to measure the depths, gradients, pressures, and flow characteristics of subsurface water in soils and fractured rock. The simplest procedures are those which deal with groundwater elevations and gradients in unconfined conditions; these properties ordinarily are measured in open pipes. Permeability , or hydraulic conductivity, measurements require more complex equipment and procedures, as described below.

Permeability of soil masses ordinarily is the result of flow through the interstitial voids between the particles that make up the soil mass. Flow through rock masses most commonly occurs through interconnected systems of fractures or other discontinuities, rather than through interstices between the particles that make up the rock mass. However, some rocks - most particularly weakly cemented or compaction sandstones - may have an intergranular permeability greater than that which is initially sought at the outset of a grouting program. Solution cavities in limestone, dolomite and evaporite rocks present special conditions of high permeability, as do lava tubes and shrinkage cracks in certain extrusive igneous rocks.

The permeability of soil and rock can be determined from tests performed on samples in the laboratory or from tests performed in situ. The type of test selected, and the location of the test depends on the type of soil, required accuracy of the results, and cost of performing the test. The following is a brief summary of the laboratory and field tests commonly used for determining permeability. For greater detail on test procedure, the reader is directed to many of the excellent text books and manuals available, some of which are included in the references.

Regardless of the soil or rock type, most permeability tests are based on Darcy's Law:

$$Q = kiAt$$

where:

- Q = quantity of flow over a given area
- A = area normal to the direction of flow
- i = hydraulic gradient
- t = length of time
- k = coefficient of permeability

Table 5.1 provides an overview of soil types and methods of determining permeability.

5.2 Laboratory Methods - Direct

5.2.1 Constant Head Permeability Test

This is one of the more common, economical, and easiest permeability tests to perform. Its use is limited to relatively permeable soils such as gravel, sand, and clean silt. For more impermeable soils, the length of time needed to conduct the test becomes excessively long. The test is performed by passing a continuous supply of water, under a constant total head, through a chamber of constant cross-sectional area containing the soil sample. The quantity of water is collected over a given time period, and permeability is calculated from Darcy's Law.

Permeability and Drainage Characteristics of Soils[*]

<table>
<tr><th></th><th colspan="11">Coefficient of Permeability k in cm per sec (log scale)
10^{2} 10^{1} 1.0 10^{-1} 10^{-2} 10^{-3} 10^{-4} 10^{-5} 10^{-6} 10^{-7} 10^{-8} 10^{-9}</th></tr>
<tr><td>Drainage</td><td colspan="6">Good</td><td colspan="2">Poor</td><td colspan="3">Practically Impervious</td></tr>
<tr><td rowspan="2">Soil types</td><td colspan="2" rowspan="2">Clean gravel</td><td colspan="3">Clean sands, clean sand and gravel mixtures</td><td colspan="4">Very fine sands, organic and inorganic silts, mixtures of sand silt and clay, glacial till, stratified clay deposits, etc.</td><td colspan="2" rowspan="2">"Impervious" soils, e.g., homogeneous clays below zone of weathering</td></tr>
<tr><td colspan="2"></td><td colspan="5">"Impervious" soils modified by effects of vegetation and weathering</td></tr>
<tr><td rowspan="2">Direct determination of k</td><td colspan="6">Direct testing of soil in its original position—pumping tests. Reliable if properly conducted. Considerable experience required</td><td colspan="5"></td></tr>
<tr><td colspan="5">Constant-head permeameter. Little experience required</td><td colspan="6"></td></tr>
<tr><td rowspan="2">Indirect determination of k</td><td colspan="2"></td><td colspan="3">Falling-head permeameter. Reliable. Little experience required</td><td colspan="3">Falling-head permeameter. Unreliable. Much experience required</td><td colspan="3">Falling-head permeameter. Fairly reliable. Considerable experience necessary</td></tr>
<tr><td colspan="5">Computation from grain-size distribution. Applicable only to clean cohesionless sands and gravels</td><td colspan="4"></td><td colspan="2">Computation based on results of consolidation tests. Reliable. Considerable experience required</td></tr>
</table>

[*] After Casagrande and Fadum (1940).

Table 5.1 Permeability and Drainage Characteristics of Soils
from Soil Mechanics in Engineering Practice, Terzaghi and Peck (1967).
Reprinted by Permission John Wiley & Sons, Inc.

5.2.2 Falling Head Permeability Test

The falling head permeability test is appropriate for use on soils with moderate to low permeability, such as silty and clayey sands and gravels, clay silt and clays. For this test, the soil sample is contained in a cylinder of constant cross-sectional area. Water enters the sample through a small diameter tube,or standpipe, at the top of the cylinder, and exits at the bottom of the cylinder. The quantity of water that passes through the sample over a given period of time is determined by computing the volume from the change in head in the standpipe. Since low permeability soils pass small quantities of water, the diameter of the standpipe (and, therefore, the volume), must be sized appropriately for the type of soil being tested. An appropriate form of Darcy's Law, which accounts for the decreasing gradient with time, is used to calculate permeability (Cedergren, 1967).

5.3 Laboratory Methods - Indirect

Over the past 40 years, many types of permeability tests have been performed on a wide variety of soil types. Correlations between grain size, void ratio and density have been established to estimate permeability for a given soil when actual test data is unavailable, or uneconomical to obtain.

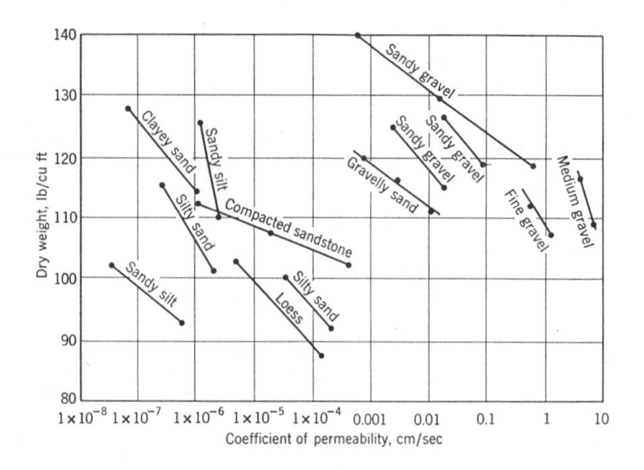

Figure 5.3.1 Correlation of Permeability to Unit Weight
from Seepage Drainage and Flow Nets, Cedergren (1967)
Reprinted by Permission John Wiley & Sons, Inc.

For preliminary design work and conceptual studies, empirical approaches are often useful and economical. One typical empirical chart is shown in Figure 5.3.1.

5.4 Field Methods

The permeability of a soil or rock deposit can vary widely even within a zone of similar material. For example, small lenses of gravel in a sand deposit will

allow much greater quantities of water to flow than through the thicker sand deposit. Also, in natural deposits, the coefficient of permeability in the horizontal direction is usually greater than that in the vertical direction. For these reasons, it is usually more advantageous to determine the coefficient of permeability from in situ field tests rather than from tests on laboratory samples.

There are several types of well pumping tests that are commonly used. The simplest of these tests requires one well. More complicated tests require at least one test well and one or more observation wells or piezometers. Due to their relatively high cost, field tests are usually used on large projects. A brief discussion of several field permeability tests follows. Because these tests require considerable experience to conduct and to evaluate the data, the reader is directed to the references include in this section for more detail on these tests.

5.4.1 Bore Hole Test - USBR Designation E-18 US BuRec (1974)

Bore hole tests are some of the least costly field test methods and were developed by the USBR to economically determine in-place permeability. These are pump-in tests and can be performed in two ways.

a) Open End Test - For this test, an open ended casing is drilled or driven to the stratum to be tested. The casing is carefully cleaned out without disturbing the soil at the end of the casing. The test is conducted by maintaining a constant head by adding water to the pipe through flow meter. When necessary, water can be pumped in under constant pressure. Permeability is calculated from the applied head and the rate of infiltration.

b) Packer Test - This test is similar to the open end test except that one or more packers are placed in the bore hole. For cased holes in soil, a single packer is placed at the end of the casing. Where a stable uncased hole can be made in rock, one or more packers can be placed within the bore hole itself. Water under pressure and known flow rate is pumped into the hole below a single packer or between two packers in an uncased hole.

Water pressure tests utilizing packers (packer tests) are commonly performed in borings with stable sides, i.e. typically in rock, to assess the permeability in the vicinity of the hole before grouting. Packer tests ordinarily are performed by the grouting contractor's crew, with the data being recorded by trained and experienced technical personnel. Optimally, so as to minimize the likelihood that conditions in the hole may change prior to grouting, these tests should be performed shortly before grouting the hole (Weaver, 1991). Experience is required for the evaluation of the significance of packer test data. Houlsby (1990) presents extensive background information on the use of packer tests, together with advice on procedures for conducting and interpreting them.

5.4.2 Pumped Wells with Observation Piezometers

Another, but more costly method of determining in situ permeability, is with a well pumping test. This test can be conducted with steady state or non steady state conditions and is usually performed by pumping water from one central well while water level readings are taken in two or more surrounding wells. As water is pumped from the well, the water table in the surrounding soil is lowered and slopes toward the well. The flow rate, depth to the water surface, distance from the pumped well to the observation wells, are recorded. Permeability is calculated from formulas derived from Darcy's Law (Cedergren, 1967).

This type of test measures average permeability over a very large area compared to laboratory tests. It is usually used for determining the permeability in cohesionless deposits beneath dams or containment facilities. The test requires special equipment and a significant amount of time (several days to several weeks depending on the purpose of the test) to complete. Because of the time and expense of performing these tests, a thorough site investigation should be considered to limit the number of tests needed.

5.4.3 Weir

A weir is a simple, indirect method for determining the permeability through or under a formation, where the discharge is concentrated and collected in an open channel. In its simplest form, a weir is a small dam where water flows through either a rectangular or V-notch opening of known geometry. By measuring the depth of the depth of the water flowing over the weir, the flow rate can be determined from appropriate formulas (Dougherty and Franzini, 1965).

Weirs do not provide direct readings of permeability; they do provide accurate flow rate data from which the permeability can be approximated, or changes in flow can readily be determined.

6. Other Methods

6.1 Pulse Echo

The pulse echo method is a special application of ultrasonic methods. It uses an ultrasonic pulse that is reflected from discontinuities. This approach has long been used for the evaluation of metal and for weld inspection. The method is requires a special device that both generates a pulse and receives the reflected pulse. The travel time between the initial signal and the arrival of its reflection is directly related to the thickness and wave speed within the medium. One application of this method is in looking for voids in the soil outside of a pipe or tunnel. Where soil is in tight contact with the pipe or tunnel lining, most of the pulse's energy travels across the contact between the concrete and the soil, resulting in a low amplitude or nonexistent reflection. Where there is a void, most of the energy of the pulse is reflected and a large amplitude return pulse is obtained. Grouted voids will not produce reflections.

PART D SELECTION OF VERIFICATION METHODS

1. Introduction

When selecting methods of verification, it is important to consider the importance of the parameter being measured, required accuracy, site limitations, method limitations and other factors. Sometimes it may be necessary to use one or more indirect methods to obtain a parameter. Consideration should be given to using a large quantity of inexpensive indirect or limited accuracy tests with a few confirmatory direct tests. For long term applications, the stability and durability of the instruments will be an important factor. Accessibility and ease of maintenance should also be considered, especially for the long term. The cost of instrumentation should consider the cost of installation, maintenance, monitoring, interpretation of the data, and the possible loss of a number of instruments due to damage by construction equipment, vandalism, etc.

Some factors relevant to method selection are discussed in this section. The discussion is divided into four section based on the goal of the monitoring. These are: detecting grout, measuring changes in permeability, density, strength and modulus, and measuring movement

2. Detecting Grout

One of the simplest performance measures is simply determining whether grout has been placed in the desired locations. Commonly a test section may be grouted and excavated to verify the effectiveness of the technique prior to full scale production grouting. Verification of the test section may involve several different methods including excavation, bulk sampling, coring, field tests such as the use of phenolphthalein for alkaline grout and penetration testing.

2.1 Grout in Voids

2.1.1 Voids Adjacent to Structures

Where grouting is used to fill voids beneath slabs or behind structures, a simple method such as tapping on the concrete surface with a metal rod and listening for a hollow sound may be sufficient. For more complex or thicker structures a non-destructive method such as pulse echo or ground penetrating radar may be appropriate. Pulse echo would indicate the presence of voids behind the walls of a pipe or tunnel lining. Both pulse echo and GPR would give more or less continuous readout of slab support conditions. Destructive methods such as coring or saw cutting are readily available but require repair of the structure and can be messy in occupied spaces. The destructive methods may be used to explore problem areas discovered by the indirect pulse echo or GPR procedures. Colin (1982) has used resistivity monitors and nuclear density gages to verify annular grouting of piles for offshore structures.

2.1.2 Deep Voids

Deeper voids, such as those in karst, may be more difficult to verify, unless their locations are clearly known in advance. In many cases, the grouting process itself is the best detection and verification technique. For example, if an initial injection hole encounters a void and has a substantial grout take, redrilling at that location should indicate that the void has been filled with grout and reinjection should result in a low take if the initial grouting was effective. For large areas with frequent voids, geophysical methods may detect an average increase in the shear wave velocity after grouting which is indicative of grout filling the voids. For random voids, the likelihood of randomly (or regularly) spaced core borings hitting voids is quite low.

2.2 Permeated Grout

Where grout has permeated the soil structure, detecting its presence can be a challenge. Where the soil has a rigid structure such as for gravel or fragmented rock, the methods used to penetrate the soil matrix may destroy the grout structure or wash away the grout. Where the soil is fine grained or stratified, the degree of permeation may be variable or very small.

2.2.1 Coarse Materials

The presence of grout in coarse grained materials or fractured rock can be directly evaluated by destructive methods such as test pits and core samples. Some grouts can be easily identified visually, but others may need chemical enhancement by dyes, or pH detectors such as phenolpthalene. Sometimes soil cement grouts are difficult to identify because they closely match the color and consistency of the soil matrix. However, the distinct odor of Portland cement can assist in identifying the grout.

Coring is not practical for low strength grouts in a coarse matrix. Coring is typically only successful where the grout strength is high enough to withstand the coring process. For the same reason, hand excavation may be needed in test pits to identify the grout. The use of dyes added to the grout may to aid visual identification.

Indirect methods such as penetration tests, and many geophysical methods may not be sensitive enough to detect very small changes in soil stiffness produced by grouting in coarse soils. These methods may be appropriate where a very stiff/strong grout is used and significant changes in soil stiffness are expected.

2.2.2 Sands and Fine Soils

Where grout permeates medium to fine grained soils, significant changes in the soil strength may occur that will aid verification. Direct methods such as excavating test pits are effective but often require supplemental detection aids such as dyes added to the grout or pH detectors such as phenolpthalene. Conventional core sampling is difficult, primarily because the matrix is eroded during the drilling process. Other methods of probing may in many cases distinguish between the stiffer grouted soils and ungrouted soils. Standard penetration tests (SPT) may obtain disturbed samples that can be inspected for the presence of grout. SPT will not usually obtain an intact sample, but the method is relatively inexpensive and commonly available. The average increase in SPT blow counts may be used, but individual blow counts may not be representative.

Core samples of grouted rock, and in some special cases grouted soil, provide a direct means of determining the effectiveness of grouting. The sample can be examined visually or can be examined under the microscope to determine the presence of grout and the effectiveness of the grout in permeating the pore spaces and fissures. This method has a distinct advantage of providing a physical sample to view and evaluate. Positive evidence of the presence of the grout can be readily determined. One disadvantage of this method is that it is difficult and costly to obtain samples of the grouted material (Munfakh, 1991). The success of obtaining an intact representative sample depends on several factors. These are a) the strength of the grout, b) the strength of the formation, c) the type of material, i.e., rock or soil, and d) the drilling technique used. Coring the formation provides a bore hole which may also be used to conduct water pressure tests. In addition, the core sample can be used correlate the visual structure with the water pressure tests results.

Indirect approaches will generally be most suited to detecting grout permeation of medium to fine grained soils. Since the stiffness of the soil can be expected to change significantly, methods that measure stiffness will likely be appropriate to detect grout. Methods such as geophysics, PMTand in some cases CPT and SPT, should be considered. Appropriate geophysical methods include seismic methods and in some cases resistivity and ground penetrating radar.

2.3 Grout Inclusions

Many grouting methods produce a grout inclusion within the soil matrix, the shape of the inclusion affects the selection of verification method. Compaction grouting produces typically regularly shaped balls of grout, fracture grouting produces thin lenses of grout and jet grouting and soil mixing produce cylinders or parts of cylinders of grouted soil. The primary reason to verify inclusions is when they are used to form structural elements or seepage barriers. However it may be important to locate the extent of compaction or fracture grouting to verify that it has been correctly placed (e.g. in settlement control applications).

2.3.1 Continuous Columns

The continuity of grouted elements can be verified directly by drilling or coring through the center of the hardened grout structure, however, where grouting has been done under existing foundations, this is often not practical. Load tests can also be used to directly evaluate the load carrying capacity of grouted columns. In cases where the grouted elements provide shoring or underpinning, their integrity can often be observed directly when the grout is exposed during excavation.

Indirect methods may be useful in some applications. Seismic methods can detect the continuity of grouted columns. This is a special application of the downhole method, where a geophone is placed in a borehole immediately adjacent to the grouted column or is cast into the bottom of the column. The amplitude and travel time of the seismic pulse are used to evaluate the continuity of the grout. This is not a commonly used or widely accepted method, but has had application in practice.

Large caliper logs (umbrellas) have also been utilized to measure diameters of liquid jet grouted columns. However, in sandy soils the wet soil and grout mixture may be too dense to install the umbrella beyond shallow depths, the umbrella is difficult to open and the evaluation of column size which is made

on the basis of resistance to opening becomes a difficult, qualitative assessment.

2.3.2 Other Inclusions

The location and presence of grout inclusions can be evaluated by excavating test pits and in many cases by simple probing. Extensive use of test pits have been used to evaluate alternative grouting methods after test grouting, prior to final design (Atwood and Lambrechts, 1995). Simple probing methods can be cost effective and easy to interpret in soft soils. Where drilling equipment is needed to penetrate the grouted matrix, such as in gravel or rocky soil, the inclusions can be difficult to distinguish from other materials.

Indirect methods such as GPR can under the right conditions identify the location and extent of grout in the soil. Tomographic analysis of this and other geophysical methods can produce a contour map of soil stiffness, conductivity, or electromagnetic transmissibility that can be use to identify the presence of grout. Tomography requires large volumes of data, usually from boreholes, and can be too expensive for many projects.

3. Evaluating Changes in Permeability

3.1 Introduction

Grouting is frequently used to reduce the permeability of rock and soil. Part C 5.0, above, discusses many of the commonly used direct and indirect methods of determining permeability in soil or rock. This section reviews the application of the various test methods to verify the effectiveness of grouting performed to reduce permeability or hydrostatic pressure.

The type of test that is selected depends on several factors. These are:

a) Type of formation
b) Cost
c) Required accuracy of data
d) Experience of testing personnel
e) Availability of testing equipment

Once the test method has been selected, the obvious technique is to compare the pre-grouted permeability with the post-grouted permeability. In reality, such a simple comparison can not always be made and considerable engineering judgement and interpretation of uncorrelated data are required.

For example, the permeability of an alluvial deposit beneath a dam is determined prior to construction of the dam with an extensive well pumping test. The well test indicates that a grout curtain is required to reduce seepage beneath the dam. However, after construction of the dam and grout curtain, access to the test site is no longer possible due to the presence of the dam and reservoir. For this situation, one would have to depend on data from piezometers, discharge weirs, or other methods to indirectly determine the effectiveness of the grout curtain. For an existing dam, however, the effectiveness of the treatment can be clearly demonstrated (e.g., Riemer, et al., 1995, Bruce, et al., 1993).

3.2 Direct Methods for Permeability Determination

Direct methods of testing permeability for the purpose of verification of the effectiveness of grouting provide a determination of the permeability of the soil or rock based on known hydraulic gradients and measured volumes of seepage. These methods are as follows:

3.2.1 Laboratory Testing

Use of the falling head or constant head permeability tests can be used to determine the permeability of samples of grouted versus un-grouted soil. This test provides direct comparison of the effectiveness of grout on permeability. One of the most significant limitations of using this type of test is that the permeability test is only performed on a small discrete sample. One has to assume that the sample is representative of the entire grouted formation. Another limitation in using this type of laboratory test is that it is often difficult, to obtain an undisturbed sample of the grouted soil. For both chemical and cement grouting, the use of rotary boring techniques are often too aggressive and may disturb or completely destroy the sample.

Bench tests can be performed on soil samples that have been injected in the laboratory. This may be useful in planning the grouting program and evaluating problems with grout permeation and performance, particularly where unanticipated soil types are encountered during construction.

Falling head and constant head permeability tests on grouted samples have found extensive use in laboratory research on new grouts or in determining the engineering properties of grouted soils. Krizek and Helal (1992) and Siwula and Krizek (1992) have used such tests to study the effects of sodium silicate grout and ultrafine cement grout on granular materials.

3.2.2 Bore Hole Testing

There are several methods of determining the permeability, and therefore the effectiveness of grouting, from boreholes drilled into soil or rock. These tests vary in their complexity and in the accuracy of the results obtained.

3.2.2a USBR Bore Hole Test, Designation E-18

This test is primarily used on granular soils and at relatively shallow depth. This test is most commonly used to determine quickly and economically the permeability of soils close to the surface. However, this test can also be performed in a grouted soil assuming that a bore hole can be drilled or advanced into the grouted soil. One limitation of this test is that only the permeability of the grouted soil at the end of the bore hole is being tested. This represents a rather small volume of the entire grouted formation. Also, the permeability readings provided by this test tend to decrease over time as a filter cake of fine soil tends to accumulate at the bottom of the bore hole, thus decreasing the measured permeability at that location. This test is also limited in that it provides an estimate of permeability only within a given stratum, i.e., that stratum in which the boring terminates.

3.2.2b Packer Testing in Grout Holes

Water pressure tests utilizing packers are commonly performed in borings with stable sides, i.e., typically in rock to determine the relative permeability of the formation before and after grouting. Evaluating the permeability of the formation prior to grouting will provide some indication as to the type of grout mix that is required and the approximate grout take that can be anticipated. When these tests are performed after grouting, they will show the effect of grout permeation from within the grouted hole or the effect of grout from adjacent holes. The most common unit of measure for reporting the result of a grout hole water pressure test is the Lugeon unit. One Lugeon unit equals one liter per meter of bore hole depth per minute at excess head of I Mpa (145 psi). For comparative purposes, one lugeon unit is equivalent to a permeability of 1.3×10^{-5} cm/sec (10' ft per year) . Because this test is relatively simple and economical to perform it can be conducted frequently in both first stage, second stage and third stage grouting, thus providing a very reliable means of comparing the reduction in permeability within a grouted rock formation.

The advantages of bore hole testing using Lugeon units is that the same equipment that is used to perform this test is also used for performing grouting, therefore, economies for this test are realized. The test can also be performed by the grouting crew, i.e., with semi-skilled or unskilled labor under the observation of an experienced technician or engineer. Experience is required in evaluating the data to obtain the proper conclusion and assess the effectiveness of grouting when the test is performed after grouting. The technique for conducting this test, interpreting the results and extensive background on the use of the test are discussed most thoroughly by Houlsby (1990).

The data obtained from packer tests, together with a geologic interpretation of subsurface conditions, may be used to select the appropriate grouting material and initial grout formulations.

It is essential to recognize that packer testing provides the only reasonable means of assessing whether or not initially open fractures intersected by adjacent grout holes have been satisfactorily filled with grout. Weaver (1991) lists 10 possible conditions other than complete grouting that might explain low grout takes or apparent "refusal."

3.2.3 Weirs

Weirs are most commonly used to determine the flow rate in open channels. Weirs find their application in geotechnical practice in determining the flow from the downstream faces of earth or rock dams or flow through natural formations that form the sides and containment of reservoirs. By monitoring the flow through weirs one can easily compare the quantity of flow prior to grouting with the reduction of flow after a dam abutment or foundation has been grouted. A typical application of using weirs to determine effectiveness of grouting is as follows: naturally occurring fissures or solution cavities often can provide a leakage path for water from a reservoir to a downstream location. A weir is located at the point where the water exits the ground to determine the pre-grouting flow rate. The flow through the weir is monitored over time as grout is placed beneath the dam or in the abutment. One of the advantages of this method is that results are obtained immediately and the effectiveness of grouting can visually be observed. It is also relatively inexpensive to set up a weir and monitor the flow, particularly with remote sensing equipment. One of the main drawbacks of using this type of test is that it only provides information on the quantity of flow at the discharge point. It provides no information as to where the water is entering or traveling through the soil or rock formation.

3.2.4 Dye Tests

This procedure does not provide a quantitative determination of permeability of a formation. However, dye tests can be used to evaluate the seepage path of water through a naturally occurring formation or man made structure. For example, dye is injected at a discreet location upstream and the various downstream discharge points are monitored for the appearance of the dye. Dyes have been effective to identify open seepage channels for chemical grouting (Karol, 1982). Chemical tracers, both injected and natural have been effective in identifying spots in a grout curtain where additional grouting is needed (Riemer, et al., 1995). This test is very economical to perform and provides readily understood qualitative results. Some of the disadvantages are that it may be difficult to locate the source of the leak with dye and multiple trials and injections of the dye may be required. Also where the rate of flow is very high, large quantities of dye may be required due to the rapid dilution of dye concentraion. Finally, where the rate of flow is very low, it may take an excessively long time for the dye to travel from the entrance point to the discharge point.

3.3 Indirect Methods for Permeability Determination

Indirect methods are those that provide an indication of the effectiveness of grouting without actually testing for the permeability of the soil or rock.

3.3.1 Grouting Records

The methods of grouting to reduce permeability are discussed and reviewed in detail elsewhere in this document. These methods are not repeated in detail here. Because grouting can be a relatively costly procedure, information obtained during every stage of the grouting operation is valuable and can be used to help determine a) the appropriate type and consistency of the grout, and b) the effectiveness of the grouting operation. Much valuable information is obtained during the drilling and grouting stages. During the drilling stage very useful information on the formation can be obtained from soil samples or core samples where the porosity and the fracture pattern of the soil or rock is readily observed. Also by observing the quantity of drill water that has been returned from the bottom of the hole, a preliminary estimation of the permeability of a jointed rock mass can be obtained. Once the bore hole is complete, pressure testing as previously described can be performed.

During the grouting operation, the grout logs provide information on the pressure that the grout was injected under, and the volume of grout that was injected. As the primary grouting proceeds, a pattern of grout flow and pressure is established. The same data are recorded for the second and third stages of grouting. One would anticipate that the pressures should increase for the latter stages of grouting and the volume of injected grout should decrease. Careful and thoughtful evaluation of the grout logs provides one of the most efficient and effective methods of determining the effectiveness of grouting. The grout records provide real time information on the permeability of the formation with respect to the grout. Frequently, grouting records provide enough information such that the need for other direct or indirect verification are eliminated. It has been suggested that once a given hole or series of holes reaches refusal, i.e., it can no longer accept grout under any pressure or a predetermined limiting pressure is reached, it can be assumed that all fissures or voids have been satisfactorily filled with grout. However, it is possible, depending upon specific site geology and grouting conditions, that the unfilled fissures or voids may remain between

between injection locations. For this reason, additional verification is often required.

It is essential that detailed and complete records of all aspects of grouting operations be kept. These data while eventually forming a potentially valuable part of the "as-built" construction records for the project, are even more important as a basis for: 1) continually reinterpreting the subsurface conditions, 2) for interpreting the apparent changes in those conditions as a result of grouting, 3) as a basis for assessing whether or not further grouting is needed when the planned maximum hole spacing in the split-spacing sequence is reached, and 4) for identifying a possible need for modification of procedures for optimizing the results. Additionally, the records are used as a basis for payment and for adjudicating claims. Examples of grouting records typically needed for recording data from dam foundation grouting operations are presented in an ASCE publication by Weaver (1991). Millet and Engelhardt (1982) present a rational method for evaluating structural grouting in rock using grouting records. Lamb and Hourihan (1995) present an excellent case history of split spaced compaction grout injections used to verify the effectiveness of the production grouting.

Optimally, the drilling data for each hole will include drilling rate, unusual action of the drill (such as "chattering", rods dropping, etc.), color and clarity of water return, nature of drill cuttings, and gain or loss of drill water. If the records indicate that drilling continued after partial or total loss of water circulation in fractured rock, it ordinarily can be inferred that potentially groutable openings may have been clogged with cuttings and that the hole should be replaced.

Packer test data should be presented on a pressure grouting log for that same individual stage and hole, not only as a guide to selection of the initial grout mix but as an indication of whether or not a significant grout take should be expected. (Low grout takes coupled with packer test data above the target standard may serve as an indication that the specified "limiting grouting pressure" may be too low for adequate grouting to be effectively accomplished.) The data to be entered will include: 1) components of the grout mix, 2) grouting pressure, 3) rate of injection, and 4) measurements (direct or indirect) of volume of grout take. Periodic entries on the pressure grouting log commonly indicate a gradually decreasing rate of take as injection of grout proceeds. However, other trends may be noted, and may require corrective action. Information on interpretation of pressure-take trends is presented by Weaver (1991, Table 6).

Confident interpretation of drilling, packer testing and grouting data requires that these data be continually plotted, on a daily basis, at an equal horizontal and vertical scale on a cross section profile along the grout curtain. Ideally, this section also will include interpretive projections of geologic defects observed at the foundation surface. Only in this way can the interrelationships of the various types of data be placed in a proper context.

3.3.2 Monitoring Wells and Piezometers

Monitoring wells and piezometers provide a means for determining the ground water location or piezometric head within a earth embankment or aquifer. Under natural or normal conditions and steady state flow, the piezometric head at a given location remains constant. In hydraulic structures, such as dams and aqueducts, water pressure within the surrounding formation may be detrimental to the satisfactory performance of that structure. For this reason, grouting may be implemented to reduce the hydrostatic pressure. Piezometers and observations wells provide a direct indication of the hydrostatic head at that location. By comparing piezometer levels before grouting with those during and after grouting, the change in head can be determined. This provides an indirect means of observing the effectiveness of grouting. This data can be utilized with flow net diagrams and calculations made to determine the quantity of flow or uplift pressures beneath the structure. Pneumatic piezometers, and standpipes have also been used to measure the excess pore pressures generated by jet grouting (Ho, 1995).

4. Evaluating Changes Density

Compaction grouting may be done to increase the density of the soil. Density changes can be measured by comparing the in situ density before and after grouting. The density testing method and density standard should be consistent with the expected range of improvement from the grouting method used. Compaction grouting increases the density of the soil surrounding the bulb of grout injected. The soil density between grout bulbs will not be uniform.

Direct Methods of density include the Sand Cone test, Rubber-Balloon test, Drive Cylinder test and for coarse soils the Sleeve Method test. Other direct methods such as weighing cut block samples and the like may also be feasible depending upon the nature and cohesive strength of the grouted soil. All of these methods except possibly the Drive Cylinder test can only be performed on the ground surface or in a test pit. The most difficult aspect of density testing below the ground surface is obtaining an undisturbed volume of soil. Direct methods for density testing at depth are only appropriate for soils that possess sufficient cohesion to hold together during sampling.

Indirect Methods for evaluating soil density include nuclear density test, Cone Penetration (CPT), Dilatometer (DMT) and Pressuremeter (PMT) Tests, and Seismic Methods. Seismic methods have been shown to be effective in measuring changes in soil density where inclusions make up a relatively small proportion of the soil volume, but should be correlated with physical tests (Byle, et al., 1991). Partos, et al., (1982) found the downhole nuclear density probe to provide repeatable testing at the same locations during the grouting that could not be achieved with SPT or CPT.

Both CPT and SPT have been used to evaluate densification by compaction grouting. SPT has been used successfully for measuring densification by compaction grouting (Salley, et al., 1987, Tokaro, et al., 1982, Welsh 1986). Density is commonly related to penetration resistance for natural soils. However, relative density correlations from penetration resistance tests will generally not be valid for the remolded or grout permeated soil after grouting; so unless a new site specific correlation is developed the penetration tests may not be a good measure of density improvement. Welsh (1986) indicates that typically a three- to five-fold improvement in the SPT N value can be obtained up to N=25 and CPT value increases from 8 to 15 Mpa are reported.

The nuclear density test when properly calibrated, generally gives reliable measurements of soil density. The conventional nuclear testing gage can be used to measure density of the upper 200 to 450 mm of the soil. Down-hole versions of the nuclear densimeter probes can be used at any depth in a borehole to measure density below grade within 100 to 200 mm of the borehole.

5. Evaluating Changes in Strength and Modulus

5.1 Load Tests

In-situ plate load tests can be made on grouted and ungrouted soils by applying incremental loads on a steel plate placed on top of a leveled soil surface. This is a direct modulus measurement and can be used to measure strength if taken to failure. Testing can be performed at different elevations and locations, however the costs of multi tests can be high and warranted only on a complex and critical grout program. Other considerations for the use of plate load tests may include:

- Labor costs in preparing and conducting such tests.
- Engineering fees
- Results limited to location and depth of tests
- Stratified or non-homogeneous soils require a large number of tests

Davidson and Perez (1982) report measurements of a 3 to 10 fold increase in modulus and 20 to 200 percent increase in chemically grouted ultimate stress using plate load test on sands.

Pile load tests on similar approaches can be used to measure the capacity of grouted structural elements. Static load tests have been successfully used to measure the improved capacity of caissons after finely ground cement grouting (Bruce, 1995).

5.2 Laboratory Strength Tests

Laboratory shear tests also provide a direct strength measurement. These tests require that block or core samples of grouted soils be obtained from the site and be brought into the lab for testing. The blocks can be trimmed for direct shear, unconfined compression or triaxial compression tests. Likewise these tests can be performed on the cored samples if sampling disturbance is not excessive.

In the field, large diameter cores, 15 cm or larger can be taken of grouted soils at various depths and locations for lab testing. However, where large grain soils such as gravels and cobbles are present poor core recovery can be experienced, especially in chemically grouted soils. Water and other liquid drilling fluids can damage samples and air is often used to aid in coring.

Reconstituted samples can be prepared using the same grout mix and field soil samples for compression testing. However, the similarity between these tests results and those from field trimmed block samples will depend the uniformity of in situ grout permeation and homogeneity of the soil mass. In relatively clean sands the results would be expected to be much more similar than in a more heterogeneous deposit with varying fines content.

Factors to consider for use of laboratory shear tests are:

- Feasibility of obtaining block samples or large diameter cored samples of grouted soils from a representative number of depths and locations.
- Selection of sampling techniques for best sample recovery (consider soil and grout properties and presence of gravels and cobbles or other large particles that could affect sample quality).
- Disruption of the site during sampling
- Time required for laboratory preparation and testing of samples.
- Restoration of the site after sampling

Except for projects that routinely expose grouted soils, such as tunnelling, the number of tests that can be economically performed is relatively limited. As stated earlier, in many cases, the project would have to be critical and the soil profile complex to justify direct testing.

5.3 Pressuremeter Tests

Pressuremeter (PMT) and dilatometer (DMT) tests can measure the change in soil modulus due to densification. The relative performance of these methods has been reported by Welsh (1986). The DMT is limited to soils without gravel and has been used to measure modulus increases from 50 to 100 MPa (500 to 1000 tsf). Pressuremeters have been adapted to a wide variety of conditions and are capable of measuring small increases in stiffness in the range of 60 to 130 kN/m^2 (0.6 to 1.3 tsf). DMTand PMT tests provide direct measures of stiffness and indirect measures of strength but do not give data directly related to density.

In situ testing of soils can be used before and after grouting to obtain changes in the modulus of elasticity. Comparison of the moduli can be used as a measure of improvement due to grouting. PMT can also be used to study creep behavior of grouted soils where such tests are justified for loads applied to grouted soils for a long period of time (Davidson and Perez, 1982). Zero to 10 fold increases in pressuremeter modulus in chemically grouted sand (Davidson and Perez, 1982).

Drill hole preparation is critical for successful pressuremeter tests, especially in sands and soft clays (Briaud and Gambin, 1984). The prepared hole should be only slightly larger than the probe diameter. In soft or caving ground, drilling mud or synthetic drilling fluid is commonly used, however, these are generally not needed when a stable hole can be formed. The probe should be placed in the hole as soon as possible to minimize borehole deformation.

The moduli calculated from PMT compare favorably with plate load and laboratory tests and can be performed at a lower cost and at various depths, elevations and locations. Some limiting factors are difficulties associated with the coring and placement of the probe in some deposits and the relative complexity compared to other in situ testing systems such as SPT and CPT. A common single crew can perform on the order of 4 to 8 PMT's in soil in one day.

5.4 Static Cone Penetration Tests

Static cone penetration tests (CPT) are very useful to determine the degree of success of grout stabilization of soil, showing the soil strength - before and after grouting. The data obtained can, by empirical correlation, determine soil strength or modulus of elasticity. The CPT are best used in conjunction with PMT to obtain soil shear strength parameters. These CPT's can be compared to SPT's and are rapid and easy to use. Interpretation of the tests is reasonably easy.

CPT's can provide continuous data with depth or discrete data for each point tested. Rapid test rate and relatively low costs per test allow for economical comprehensive testing of a site. CPT testing is not appropriate for very dense, hard or gravelly soils or where the grout cannot be penetrated by the cone under hydraulic pressure. Increases in CPT tip resistance value due to grouting have been reported from 8 to 15 kN/m^2 (80 to 150 tsf) by Welsh (1986). A single CPT rig can perform 150 to 300 m (500 to 1000 lf) of probing per day depending on soil conditions.

5.5 Shear Wave Velocity

The shear wave velocity measured from seismic methods can be directly related to the shear and elastic moduli of the soil. Crosshole shear wave velocity has been used successfully to measure changes in modulus due to grouting (Byle, et. al, 1991; Davidson and Perez, 1982). Shear wave methods work best where relatively uniform soil conditions exist and where buried structures, boulders or other discontinuities will not interfere with the direct transmission of refracted waves. The crosshole method should be used where applications under buildings or where grouting is not of sufficient areal extent for refraction. Huck and Waller (1982) have found that surface geophysical approaches generally do not give readily quantitative results while downhole methods give more specific data for grout verification. Partos, et al., (1982) found crosshole seismic tests effective for verifying that the increase in shear wave velocity in a grouted zone met design requirements support of dynamic loads from forging equipment.

6. Detecting Movement

Method selection for detecting movement is related to the type of movement to be measured. Movement types consist of differential shear, extension/contraction, elevational and angular distortion. The selection of measurement methods is based on an evaluation of accessibility, required accuracy and response time, and cost. Dunnicliff (1988) presents an excellent discussion of the various movement monitoring devices and the basis for their use. Since there are such a wide variety of instruments, only a brief discussion of instruments common to grouting applications is included herein. The reader is directed to the literature for further information on movement monitoring (e.g. Patel, et al., 1982; Warner, 1982; Neff, 1982).

It is often desired to control or mitigate damage to structures or prevent movement between adjacent structures by grouting. Monitoring movement on flat structural surfaces is generally far less expensive than subsurface monitoring. Where the location of the movement is controlled by structural joints, cracks or other zones of weakness, measurement can be made with mechanical crack gages including pins and tape, pins and calipers, or grid crack monitors. All of these are generally inexpensive. The choice of method is made on the basis of required accuracy and span.

Where measurement is required across a gap of more than a few millimeters, pins and either tape or an extensometer are appropriate. The pins should be placed in pairs diagonally to the crack or joint to measure shear. The installation of the pins may damage architectural surfaces and this should be considered when selecting the mounting method. Pins and tapes can achieve a precision of ± 3 mm. The accuracy can be increased to ± 0.1mm by using a mechanical extensometer in lieu of the tape. However, mechanical extensometers typically cost significantly more than the tapes. Electronic extensometers or convergence gages can be used for special applications where continuous monitoring is required via a recorder or telemetry for remote monitoring.

An intermediate measure is to use grid crack monitors. These are available commercially and easily read to a precision of ± 1mm.

Extensometers should be considered where longitudinal extension or compression is expected. Typical applications might be for tunnels, to measure radial distortion due to stress relaxation, or compression of a soil layer under surcharge loading. Extensometers may be used in test sections to determine appropriate grouting procedures to control movement. Extensometers have been successfully used to measure ground movements behind a tunneling shield to aid in the design of compaction and chemical grouting to control settlements (Daugherty et al., 1995).

PART E SUMMARY AND CONCLUSIONS

There is no substitute for good engineering and a qualified contractor. Even so, conditions present under the ground surface may not be as expected. Verification testing is an essential part of achieving the desired performance of grouting. Since we cannot see through the ground, we must use various techniques to derive the information we desire from non-visual means. No single method can identify the entire picture. A combination of thorough exploration prior to design, sound engineering, quality grouting practice, monitoring and record keeping during construction and appropriate verification testing is needed to provide a good product.

The most common verification testing for strength is the simplest. Standard penetration tests are still used extensively despite their limitations. Static cone penetrometers have been used frequently and are gaining popular acceptance. Pump tests (packer tests) have seen long use in the evaluation of dam foundation grouting and are still the method of choice for evaluating permeation grouting of rock. As new applications of grouting are developed, new verification procedures are needed.

Improvements are needed in the evaluation of non-homogeneous materials. Grouting into rubble, boulders, or injecting grout bulbs into sand or fine grained soils, creates a matrix with inclusions that, while very strong and stable, can be difficult to test well. New methods are being developed now to deal with contaminated soils or ground water. Geotechnical engineers are constantly borrowing and adapting technologies from other disciplines. Miniaturization and electronics technology make continuous remote monitoring possible using telemetry. The ever increasing capacity of microcomputers makes the processing of large volumes of data easier and faster. The accuracy of instruments continually improves while the cost decreases. Undoubtedly, these technologies will change the way we think of instrumentation in the not too distant future.

APPENDIX A

TOOLS FOR SELECTION OF VERIFICATION METHODS

TABLE A1
Applicability of Selected Verification Methods by Grouting Method

TABLE A2
Applicability of Verification Methods by Grouting Purpose

TABLE A3
Applicability of Verification Methods by Soil Type

Grouting Method		Verification Method Applicability: Mechanical Methods											Chemical Methods	Geophysical Methods				Hydraulic Methods				Other Methods		
		Cone Penetration CPT	Standard Penetration SPT	Probing and Sampling	Coring	Test Pits	Flat Dilatometer	Pressuremeter	Extensometers	Settlement Plates	Density Tests	Load Tests		Seismic Methods	Acoustical Emissions	Resistivity/Conductivity	Ground Penetrating Radar	Laboratory Permeability	Borehole Permeability Test	Well Pump Test	Weirs	Pulse Echo	Downhole Nuclear Probe	Grouting Construction Records
Compaction Grout		●	●	●	○	●	●	●	◗	◗	●	◗		◗	○	○	○					○	●	●
Fracture Grout		◗	○	●	○	●	◗	○	◗	◗	○	○	○	○	◗	○	○							●
Permeation Grout	Particulate	○	○	●	◗	●		●	◗	◗		◗	●	○	◗	○	○	◗	●	●	○	○		●
	Chemical	●	◗	●	◗	●		●	◗	◗		◗	●	◗	◗	○	○	◗	●	●	○			●
Jet Grouting				●	●	●			◗	◗		◗	◗	◗		○	○	●	●	●	○			●
Soil Mixing				●	●	●			◗	◗		◗	◗	◗		○	○	●	●	●	○			●

LEGEND

- ● Generally applicable
- ◗ Sometimes applicable
- ○ Applicable only in special cases

NOTE: This table presents a summary of the general applicability of selected methods for use with the various methods of grouting. Refer to the text of this document for more details consideringt the use of these methods for specific applications.

TABLE A1
Applicability of Selected Verification Methods by Grouting Method

APPLICABILITY BY GROUTING METHOD AND PURPOSE

LEGEND:

Purpose	Generally Applicable	Sometimes Applicable	Special Applications
Improving Soil Strength	■	▣	□
Controlling Settlement	◆	⬘	◇
Reducing Permeability	▼	▽	▿
Filling Voids	⬟	⬠•	⬠

Test Method	Compaction Grout	Jet Grout	Permeation Grouts	Soil Mixing	Fracture Grouting
CPT	■ ◇	⬟	▣ ◇		▣ ◇
SPT	■ ◇	⬠•	▣ ◇		□ ◇
Probing/Sampling	■	□ ⬠			
Coring	□ ⬟	■ ⬟	▣ ⬟	■	□
Test Pits	■				
Flat Dilatometer	■ ◆				▣ ⬘
Pressuremeter	■ ◆		■ ◆		□ ◇
Extensometers	□ ◆	◆	◆	◆	◆
Settlement Plates	□ ◆	◆	◆	◆	◆
Density Tests	■ ◆				□ ⬘
Load Tests	▣ ◆	▣ ⬘	▣ ⬘	▣ ⬘	□ ◇
Dyes and Tracers		▼	▼	▼	
Seismic Methods	▣ ⬘ ⬠•	□ ◇	▣ ⬘ ⬠•	□ ◇	□ ◇
Acoustical Emissions	⬠•		⬠•		
Resistivity/ Conductivity		▿	▿	▿	
Ground Penetr. Radar	⬠•		⬠•		
Laboratory Permeability		▼	▼	▼	
Borehole Permeability		▼	▼	▼	
Passive Wells		▼	▼	▼	
Well Pump Test		▼	▼	▼	
Weirs		▿	▿	▿	
UT and Pulse Echo	⬠•		⬠•		
DH Nuclear Probe	■ ◆				□ ⬘

TABLE A2

Applicability of Verification Methods by Grouting Purpose

APPLICABILITY BY SOIL TYPE AND PURPOSE

LEGEND:

Purpose	Generally Applicable	Sometimes Applicable	Special Applications
Improving Soil Strength	■	▣	□
Controlling Settlement	◆	◈	◇
Reducing Permeability	▼	⧨	▽
Filling Voids	⬟	⯂	⬠

Test Method	Gravel	Sand	Silt	Clay	Rock
CPT		■ ◆ ▽	■ ◇	■	
SPT	□	■ ◆	■ ◇	■	
Probing/Sampling	□	■ ⬠	■ ⬠	■ ⬠	□ ⯂
Coring	□ ⬠	□ ▽ ⬠			■ ◇ ▽ ⬟
Test Pits	■ ▽ ⯂	■ ▽ ⯂	■ ▽ ⯂	■ ⯂	□
Flat Dilatometer		□ ◆	□ ◆	□ ◆	□
Pressuremeter	□ ◇	□ ◆	□ ◆	□ ◆	□ ◈
Extensometers	◈	◈	◈	◈	◆
Settlement Plates	◆	◆	◆	◆	◆
Density Tests	□ ◈	▣ ◈	▣ ◈	▣ ◈	◈
Load Tests	□ ◆	□ ◆	□ ◆	□ ◆	□ ◆
Dyes and Tracers	▽	▽	▽		▽
Seismic Methods	□ ◆ ⯂	□ ◆ ⯂	□ ◆ ⯂	□ ◆ ⯂	□ ◆ ⬟
Acoustical Emissions	⯂	⯂	⯂	⯂	⯂
Resistivity/ Conductivity		▽	▽	▽	
Ground Penetr. Radar	⯂	◇ ⯂	⯂		
Laboratory Permeability	▽	▼	▼	▼	
Borehole Permeability	⧨	▼	⧨	⧨	▼
Passive Wells	▼	▼	▼	▼	▼
Well Pump Test	▼	▼	▽		▼
Weirs	▽	▽	▽	▽	▽
UT and Pulse Echo	⬠	⬠	⬠	⬠	⬠
DH Nuclear Probe	□ ◈	□ ◈	□ ◈	◈	□

TABLE A3
Applicability of Verification Methods by Soil Type

References

Ackman T.E. and Cohen K.K. (1994), "Geophysical methods: remote techniques applied to mining-related environmental and engineering problems", *Proceedings* of the International Land Reclamation and Mine Drainage Conference and 3rd International Conference on the Abatement of Acidic Drainage, Vol 4, US Dept of Interior, Bureau of Mines Special Publication, SP 06D-94, pp 208 to 217.

AFTES(1991), "Recommendations on grouting for underground works", *Tunnelling and underground space technology*, AFTES **6**, No 4, pp 383 to 461.

ASTM *Annual Book of Standards* 1995 Volume 04.08. American Society for Testing and Materials, Philadelphia, 1995.

D 1556	Test Method for Soil In Place by Sand-Cone Method
D 1586	Method for Penetration Test and Split-Barrel Sampling of Soils
D 2166	Test Method for Unconfined Compressive Strength of Soil
D 2167	Test Method for Density and Unit Weight of Soil In Place by the Rubber Balloon Method
D 2664	Test Method for Triaxial Compressive Strength of Undrained Rock Core Specimens Without Pore Pressure Measurement
D 2850	Test Method for Unconsolidated, Undrained Compressive Strength of Cohesive Soils in Triaxial Compression
D 2922	Test Methods for Density of Soil and Soil-Aggregate In Place by Nuclear Methods (shallow depth)
D 2937	Test Method for Density of Soil In Place by the Drive Cylinder Method
D 2938	Test Method for Unconfined Compressive Strength of Intact Rock Core Specimens
D 3080	Method for Direct Shear Test of Soils Under Consolidated Drained Conditions
D 4219	Test Method for Unconfined Compressive Strength Index Test of Chemical-Grouted Soils
D 4320	Method for Laboratory Preparation of Chemically Grouted Soil Specimens for Obtaining Design Strength Parameters
D 4428/4428M	Test Method for Crosshole Seismic Testing

D 4564 Test Method for Soil In Place by the Sleeve Method

Atwood, M.J. and Lambrechts, J.R. (1995), "A Test Program to Verify Grouting Effectiveness For Boston's Central Artery/Tunnel Project", Special Publication on Verification of Geotechnical Grouting, ASCE (included in this volume) .

Baker, W.H. (1982), " Planning and Performing Structural Chemical Grouting", *Proceedings of the Conference on Grouting in Geotechnical Engineering*, W. H. Baker ed., ASCE pp 515 to 539.

Barbin Y, Kofman W, and Finkelstein M. (1993), "A geophysical radar for great depths", *Proceedings* of the 4th Tunnel Detection Symposium on Subsurface Exploration Technology, April 26-29, Golden CO.

Barker, W.H.(1982), "Planning and performing structural chemical grouting", *Proceedings of the Conference on Grouting in Geotechnical Engineering*, W. H. Baker ed., ASCE pp 359 to 377.

Benoit, J. Atwood, M.J., Findlay, R.C. and Hilliard, B.D. (1995), "Evaluation of Jetting Insertion for the Self-Boring Pressuremeter", Canadian Geotechnical Journal, 32:pp 22 to 39.

Blake W., Leighton F. and Duval W.I. (1974), "Microseismic Technics for Monitoring the Behavior of Rock Structures", *Bulletin 665*, US Dept of Interior, Bureau of Mines.

Bowders J.J., Lord A.E. and Koerner R.M. (1982), "Sensitivity study of a ground penetrating radar systems", *Geotechnical Testing Journal*, ASTM 5(3/4): pp 96 to 100.

Bratten, W.L., Bratten J.L., and Shinn J.D. (1995), "Direct Penetration Technology for Geotechnical and Environmental Site Characterization", *Geoenvironment 2000*, Y. B. Acar and D. E. Daniel eds., ASCE Geotechnical Special Publication No. 46, pp 105 to 121.

Briaud, J.L. and Gambin, M. (1984), "Suggested Practice for Drilling Boreholes for Pressuremeter Testing", Geotechnical Testing Journal, GTJODJ, Vol. 7, No. 1, pp 36 to 40.

Bruce, D.A., Luttrell, E.C. and Starnes, L.J. (1993), "Remedial Grouting Using Responsive Integration[sm] at Jocassee Dam Oconee County, South Carolina", Presented at ASDSO 10th Annual Conference, Kansas City, MO.

Bruce D.A. , Nufer, P.J. and Triplett, R.E. (1995), "Enhancement of Caisson Capacity by Micro-Fine Cement Grouting -- A Recent Case History--", Special Publication on Verification of Geotechnical Grouting, ASCE (included in this volume).

BuRec (1974), U.S. Dept of the Interior, Bureau of Reclamation, 2nd Edition, Pages 573 to 578

Byle, M.J., Blakita, P.M., and Winter, E. (1991), "Seismic Testing Methods for Evaluation of Deep Foundation Improvement by Compaction Grouting", *Deep Foundation Improvements: Design, Construction, and Testing*, ASTM STP 1089 , Melvin I. Esrig and Robert C. Bachus, Eds., American Society for Testing and Materials, Philadelphia, PA.

Cedergren, Harry R. (1967), *Seepage, Drainage and Flow Nets*, John Wiley and Sons, Inc. New York, NY, Chapter 2.

Colin, J. (1982) "Cement Grouts in Offshore Special Structures", *Proceedings of the Conference on Grouting in Geotechnical Engineering*, W. H. Baker ed., ASCE pp 843 to 846.

Daugherty, R.L. and Franzini, V.B. (1965), *Fluid Mechanics with Engineering Applications*, 6th Edition, McGraw Hill Book Co., NY, Page 268.

Daugherty, C.W., Stirbys, A.F. and Gould, J.P. (1995), "Compaction Grouting Effectiveness, A146, Los Angeles Metro Rail", Special Publication on Verification of Geotechnical Grouting, ASCE (included in this volume).

Davidson, R.R. and Perez, J.R. (1982), "Properties of Chemically Grouted Sand at Locks and Dam No. 26", *Proceedings of the Conference on Grouting in Geotechnical Engineering*, W. H. Baker ed., ASCE pp 433-449.

Dunnicliff, J. (1988), *Geotechnical Instrumentation for Monitoring Field Performance*, John Wiley & Sons, New York.

Graf, E.D.(1992), “Compaction grout, 1992”, *Grouting, Soil Improvement and Geosynthetics*, R. H. Borden ed., ASCE Geotechnical Special Publication No. 30, pp 275 to 287.

Greenhouse, J. Gudjurgis, P. and Slaine, D. (1995) " An Introduction to Applications of Surface Geophysics in Environmental Investigations", Short Course Notes, April 23, 1995, available through Environmental and Engineering Geophysical Society (EEGS), P.O. Box 1175, Englewood CO 80155, (303) 771-6101.

Hauser K., Jackson M., Lane J. and Hodges R. (1995), "Deep Tunnel Detection Using Crosshole Radar Tomography", *Proceedings*, Symposium on the Application of Geophysics to Engineering and Environmental Problems, April 23-26, Orlando, FL, Environmental & Engineering Geophysical Society, Englewood, CO.

Ho, C.E. (1995), "An Instrumented Jet Grouting Trial in Soft Marine Clay", Special Publication on Verification of Geotechnical Grouting, ASCE (included in this volume).

Houlsby, A.C. (1990), *Construction and Design of Cement Grouting: A Guide to Grouting in Rock Foundations*, John Wiley and Sons, Inc. New York, pp 61 to 62.

Huck, P.J. and Waller, M.J. (1982), "Quality Control for Grouting", *Proceedings of the Conference on Grouting in Geotechnical Engineering*, W. H. Baker ed., ASCE pp 781 to 791.

Karol, R.H. (1982), "Seepage Control with Chemical Grout", *Proceedings of the Conference on Grouting in Geotechnical Engineering*, W. H. Baker ed., ASCE pp 564 to 575.

Karol, R.H. (1982), “Chemical Grouts and Their Properties”, *Proceedings of the Conference on Grouting in Geotechnical Engineering*, W. H. Baker ed., ASCE pp 359 to 377.

Kawasaki, T. et al (1981), “Deep mixing method using cement hardening agent”, *Proceedings of 10th International Conference on Soil Mechanics and Foundation Engineering*, Stockholm, pp 721 to 724.

Koerner R.M., Reif J.S. and Burlingame M.J. (1978), "Detection methods for location of subsurface water and seepage prior to grouting", ASCE National Convention and Exposition, Chicago, IL, Oct 16-20, Preprint No 3301, 30pp.

Koerner R.M., Leaird J.D. and Welsh J.P. (1984), "Use of acoustic emissions as a non-destructive method to monitor grout", *Innovative Cement Grouting*, SP-83, ACI, Detroit MI, pp 85 to 102.

Koerner R.M., Leaird J.D. and Welsh J.P. (1985), "Acoustic emission monitoring of grout movement", *Issues in Dam Grouting*, W.H. Baker ed., ASCE Denver CO, April 30, pp 149 to 155.

Koerner R.M., Lord A.E., Bowders J.J. and Dougherty W.W. (1982), "CW Microwave location of voids beneath paved areas", *Journal of Geotechnical Engineering*, ASCE 108(GT1):pp 133 to 144.

Koerner R.M., Lord A.E. and Bowders J.J. (1981), "Utilization and assessment of a pulsed RF system to monitor subsurface liquids", *Proceedings* of the National Conference on Management of Uncontrolled Hazardous Waste Sites, Oct 28-30, Washington DC, pp 165 to 170.

Komine, H. (1992), "Estimation of Chemical Grout Void Filling by Electrical Resistivity", *Grouting, Soil Improvement and Geosynthetics*, R. H. Borden ed., ASCE Geotechnical Special Publication No. 30, pp 372 to 383.

Krizek, R.J. and Helal, M. (1992), "Anisotropic Behavior of Cement Grouted Sand", *Grouting, Soil Improvement and Geosynthetics*, R. H. Borden ed., ASCE Geotechnical Special Publication No. 30, pp 541 to 550.

Kurtz, J. (1995), "Enhanced signal processing techniques for ground penetrating radar", Symposium on the Application of Geophysics to Engineering and Environmental Problems, April 23-26, Orlando FL, Environmental & Engineering Geophysical Society, Englewood, CO.

Lamb, R.C. and Hourihan, D.T. (1995), "Compaction Grouting in a Canyon Fill", Special Publication on Verification of Geotechnical Grouting, ASCE (included in this volume).

Lord A.E. (1975), "Acoustic Emission - a Review," *Physical Acoustics*, Vol 11, Mason and Thurston eds, Academic Press, New York, NY, pp 289 to 353.

Millet, R.A. and Engelhardt, R. (1982), "Matrix Evaluation of Structural Grouting of Rock", *Proceedings of the Conference on Grouting in Geotechnical Engineering*, W. H. Baker ed., ASCE pp 753 to 768.

Munfakh, G. A. (1991), "Deep Chemical Injection for Protection of an Old Tunnel", *Deep Foundation Improvements: Design, Construction, and Testing, ASTM STP 1089*, M. I. Esrig and R. C. Bachus Eds., American Society for Testing and Materials, Philadelphia, PA.

Neff, T.L, Sager, J.W. and Griffiths, J.B. (1982), "Consolidation Grouting at Existing Navigation Lock", *Proceedings of the Conference on Grouting in Geotechnical Engineering*, W. H. Baker ed., ASCE pp 959 to 973.

Okrasinski T.A., Koerner R.M. and Lord A.E. (1979), "Dielectric constant determination of soils at L-band microwave frequencies", *Geotechnical Testing Journal*, ASTM 1(3): pp 134 to 140.

Olhoeft G.R. (1993), "Velocity, attenuation, dispersion and diffraction hole-to-hole radar processing", *Proceedings* of the 4th Tunnel Detection Symposium on Subsurface Exploration Technology, April 26-29, Golden CO.

Parish W.C., Baker W.H. and Rubright R.M. (1983), "Underpinning with Chemical Grout", *Civil Engineering*, ASCE August.

Patel, M., Miller, S.M., Li, K.L., Waugh, H. and Welsh, J.P. (1982), "Chemical Grouting for Construction of 8 Foot Diameter Tunnel Through "Little Italy", Baltimore, Maryland", *Proceedings of the Conference on Grouting in Geotechnical Engineering*, W. H. Baker ed., ASCE pp 576 to 590.

Portos, A., Woods, R.D., and Welsh, J.P. (1982), "Soil Modification for Relocation of Die Forging Operations", *Proceedings of the Conference on Grouting in Geotechnical Engineering*, W. H. Baker ed., ASCE pp 938 to 958.

Richart, F. E. Jr., Hall, J.R. Jr., and Woods, R.D. (1970), *Vibrations of Soil and Foundations*, Prentice Hall, NJ

Riemer, W., Gavard, M. and Turfan, M. (1995), "Ataturk Dam - Hydrogeological and Hydrochemical Monitoring of Grout Curtain in Karstic Rock", Special Publication on Verification of Geotechnical Grouting, ASCE (included in this volume).

Salley, J.R. et al. (1987), “Compaction Grout Test Program - West Pinopolis Dam”, *Soil Improvement - a Ten Year Update,* J. P. Welsh ed., ASCE Geotechnical Special Publication No. 12, pp 245 to 269.

Salley, R. J., Foreman, B., Baker, W. and Henry, J.F. (1987), "Compaction Grouting Test Program Pinopolis West Dan ", *Soil Improvement - a Ten Year Update,* J. P. Welsh ed., ASCE Geotechnical Special Publication No. 12, pp 245 to 269.

Siwula, J.M. and Krizek, R.J. (1992), "Permanence of Grouted Sands Exposed to Various Water Chemistries", *Grouting, Soil Improvement and Geosynthetics*, R. H. Borden ed., ASCE Geotechnical Special Publication No. 30, pp 1403 to 1419.

Task Force 27 Report, (1990), "In Situ Soil improvement Techniques", AASHTO-AGC-ARTBA Joint Committee, Washington, DC, August, pp. 291-293.

Terzaghi, K and Peck, R.B. (1967), *Soil Mechanics in Engineering Practice*, 2nd Edition, Page 177, John Wiley and Sons, Inc.

Tokoro, T., Kashima, S. and Murata, M. (1982), "Grouting Method by Using the Flash Setting Grout", *Proceedings of the Conference on Grouting in Geotechnical Engineering*, W. H. Baker ed., ASCE pp 738 to 752.

Toth, P.S.(1993), "In-Situ Soil Mixing", *Ground Improvement*, M. P. Moseley ed., pp 193 to 204.

Tsuchida, T., Kobayashi, T., Sasahara, K. and Fenner, T. (1995), " An investigation of cracks in rock slope using ground probing radar", *Proceedings*, Symposium on the Application of Geophysics to Engineering and Environmental Problems, April 23-26, Orlando, FL, Environmental & Engineering Geophysical Society, Englewood, CO.

USACE (1970), "Drained (S) Direct Shear Test", U.S. Army Corps of Engineers Engineering Manual EM 1110-2-1906, Appendix IX pp IX-1 to IX-20.

US Environmental Protection Agency (1979) *Acoustic Monitoring to Determine the Integrity of Hazardous Waste Dams*, Capsule Report, EPA/625/2-79-024, Industrial Environmental research Laboratory, Cincinnati OH.

Verfel, J. (1989). Rock Grouting and Diaphragm Wall Construction. Elsevier. New York, NY 532 pp.

Warner, J (1982), “Compaction grouting - the first thirty years”, *Proceedings of the Conference on Grouting in Geotechnical Engineering*, W. H. Baker ed., ASCE pp 694 to 707.

Warner, J. (1992),"Compaction Grout; Reology vs. Effectiveness", *Grouting, Soil Improvement and Geosynthetics*, R. H. Borden ed., ASCE Geotechnical Special Publication No. 30, pp 229 to 239.

Weaver, K.D., (1991), *Dam Foundation Grouting*, ASCE publication, New York, ISBN 0-87262-792-6, 178 pp.

Weaver, K.D. (1993), "Selection of Grout Mixes - Some Examples from U.S. Grouting Practice", *Grouting in Rock and Concrete*, R. Widmann, ed., A.A. Balkema. pp 211 to 218.

Welsh, J. P. (1986) , "In Situ Testing for Ground Modification Techniques," *Proceedings, In Situ '86*, Geotechnical Division of the American Society of Civil Engineers, Blacksburg, VA. June 23-25.

Wilson, J.F. (1968), "Flourometric Prcedures for Dye Tracing", *Techniques of Water Resources Investigations of the United States Geological Survey*, Chapter A12, USGPO, Washington, DC.

Woods, R.D., and Partos, A. (1981), "Control of Soil Improvement by Crosshole Testing", Proceedings of 10th International. Conference on Soil Mechanics and Foundation Engineering, Rotterdam.

Yen, P.T. and Gutierrez, B.J. (1993), "Evaluation of Foundation Improvement after Grouting, *Grouting in Rock and Concrete*, R. Widmann, ed., A.A. Balkema, pp 231 to 238.

A TEST PROGRAM TO VERIFY GROUTING EFFECTIVENESS FOR BOSTON'S CENTRAL ARTERY/TUNNEL PROJECT

Michael J. Atwood[1] and James R. Lambrechts[2]

ABSTRACT: A test program was undertaken to verify the effectiveness of several types of grouting that may be used during jacked tunneling under active railroad tracks for Boston's Central Artery/Tunnel Project. The program known as the Soil Stabilization Testing Program (SSTP) was carried out in soils expected to be encountered during tunnel jacking to determine grouting methods and parameters that achieve desired ground improvement. A particular emphasis was on horizontal drilling and grouting, as vertical drilling and grouting from the railroad tracks is not expected to be possible. Nearly all of the 1,150 m (3,800 lin.ft.) of drilling and 285,000 l (75,000 gal.) of grout was subsequently unearthed to qualitatively document resulting ground modification. Grouting methods included: penetration grouting with both sodium silicate and cement; and single rod jet grouting. Some grouting was oriented vertically, but the majority was oriented horizontally from within a 12 m (40 ft.) deep access excavation. Strata grouted were granular fill, organic silt, inorganic fine sandy silt, and silty clay. The sodium silicate penetration grouting was found to be effective in permeating fill voids to develop continuous grouted zones, but cement-bentonite penetration grouting was largely ineffective in permeating fill voids. Both types of penetration grouting created only grout lenses in the organic silt, inorganic fine sandy silt, and silty clay soil strata. Single rod jet grouting performed well and produced fairly consistent results in the several soil units, but encountered difficulties around wood pile and other timber obstructions. Drilling alignment for horizontal holes was largely accurate for short lengths,

---

[1] Staff Engineer, Haley & Aldrich, Inc., Cambridge, MA

[2] Vice President, Haley & Aldrich, Inc., Cambridge, MA

but considerable deviations were observed at great lengths (more than 15 m (50 ft.)) thought to be due to deflections by buried wood obstructions.

INTRODUCTION

A test program known as the Soil Stabilization Testing Program (SSTP) was undertaken during design phase of the I-90/I-93 interchange portion of Boston's Central Artery/Tunnel Project. About a mile (1.6 km) of aging, elevated expressway (I-93) in downtown Boston will be reconstructed underground, beneath the existing structure; and I-90 is being extended approximately 3 miles (4.8 km) east to Boston's Logan Airport using cut-and-cover and immersed tube tunnels (Reference 1). At the interchange, three tunnels, carrying two and three lanes each, will be jacked beneath 5 to 8 active railroad tracks that carry over 300 daily commuter and AMTRAK trains to Boston's nearby South Station. The jacked tunnels are to be installed using a compartmentalized shield, which is advanced full-face (approximately 9 m (30 ft.) high and 18 to 24 m (60 to 80 ft.) wide). The jacked tunnels are 53 to 106 m (175 to 350 ft.) long and are advanced by methods similar to those described by (Reference 2).

The SSTP has demonstrated the effectiveness of several types of grouting that may be used to reduce permeability and improve stability of ground during jacked tunneling under active railroad tracks, and has shown that jet grouting and penetration grouting can be effectively installed in various site soil strata. Grouting will be required to develop seepage barriers between the jacked tunnel construction and the surrounding ground, and to help control ground displacements beneath the tracks during tunneling. The SSTP was carried out in soils that will be encountered during tunnel jacking; the nature of these soils is discussed later.

The principal goals of the grouting test were: to document construction/grouting procedures and successes/difficulties of the grouting techniques attempted; and to visually assess results of grouting tested, thereby demonstrating the effectiveness of different grouting processes in the site soils that are adjacent to the tunnel jacking sites. In verifying that several different grouting methods reliably achieve desired ground stabilization in site soils, it was also desired to determine which of the grouting parameters attempted appear more effective in improving site soils.

It was intended to qualitatively document resulting ground modification rather than perform tests (strength, permeability, deformability) to numerically characterize the grouted ground. The visual observations of the grouted ground resulting from different grouting procedures are summarized herein in terms of diameters achieved. Controlled excavations made later to unearth the grouted ground used a backhoe for mass excavation, and hand excavation with shovels and brooms to expose and preserve the grout columns integrity and in-situ grouted character of the ground. Unearthed grout elements were visually inspected, measured and photographed as excavations progressed.

A summary of the types of grouting performed and the extent of verification of grout effectiveness by excavation is presented in Table 1. In all, the grouting program installed 12 vertical and 54 horizontal jet grout columns, 9 vertical and 40 horizontal penetration grout columns (sodium silicate and cement-bentonite type grout). The grouting methods selected for trial in the SSTP envisioned primary stabilization needs to be for reducing permeability of site granular soils and improving stability of each strata using typical, proven grouting methods/processes, i.e., permeation grout and jet grout. It was desired to test grouting methods which can be somewhat controlled to avoid movements of railroad tracks and the numerous yard switches which occur over the jacked tunnel alignments.

The horizontal grouting was installed at different elevations and into different soil types from within a sheeted access excavation, 9 m x 11 m (30 ft. by 36 ft.) in plan, that was carried to about 12 m (40 ft.) depth. The layout of the horizontal grout holes that extended out from three sides of the access excavation is shown in plan on Figure 1, and by elevation location of the three access excavation walls in Figures 2a, 2b, and 2c. Other information shown on the elevation locations is discussed later. A particular emphasis was on horizontal drilling and grouting, as vertical drilling and grouting from the railroad tracks is not expected to be permitted.

SITE AND SUBSURFACE CONDITIONS

The test site was relatively level at commencement of the program. The area had been in the tidal marsh zone during colonial times, and is known to have been traversed by a series of wharfs during the early and middle 1800's. In times since the late 1890's, other structures related to railroad activities of Boston's South Station were on or near the test site. It was expected that some of these

Table 1. Summary of Grout Installation and Verification

Type of Grouting	No. of Grout Holes	Average Length Grouted (ft.)	Soil Strata Grouted	Amount of Grouting Unearthed
Vertical Penetration Grouting				
A. Cement-Bentonite		16	Fill	100%
	4	2	Organic Silt	100%
		5 - 7	Inorganic Silt	100%
B. Sodium Silicate		16	Fill	100%
	5	2	Organic Silt	100%
		7 - 12	Inorganic Silt	100%
Vertical Jet Grouting		16	Fill	100%
	12	2	Organic Silt	100%
		17	Inorganic Silt	100%
		6	Silty Clay	100%
Horizontal Penetration Grouting				
A. Cement-Bentonite				
a. Short Holes	21	16	Fill	100%
b. Long Holes	2	62 (46 to 77)	Fill	80%
B. Sodium Silicate				
a. Short Holes	12	15	Fill	100%
b. Long Holes	5	65 (51 to 77)	Fill	80%
Horizontal Jet Grouting				
a. Short Holes	11	20	Fill	100%
	27	20	Organic Silt	100%
	6	20	Inorganic Silt	100%
	2	20	Silty Clay	100%
b. Long Holes	1	54	Fill	56%
	7	66 (52 to 74)	Organic Silt	46%

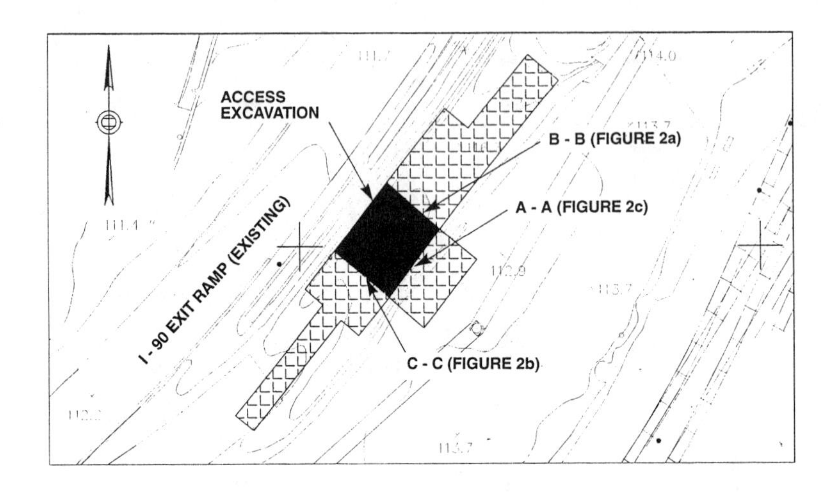

Figure 1. SSTP Site Plan

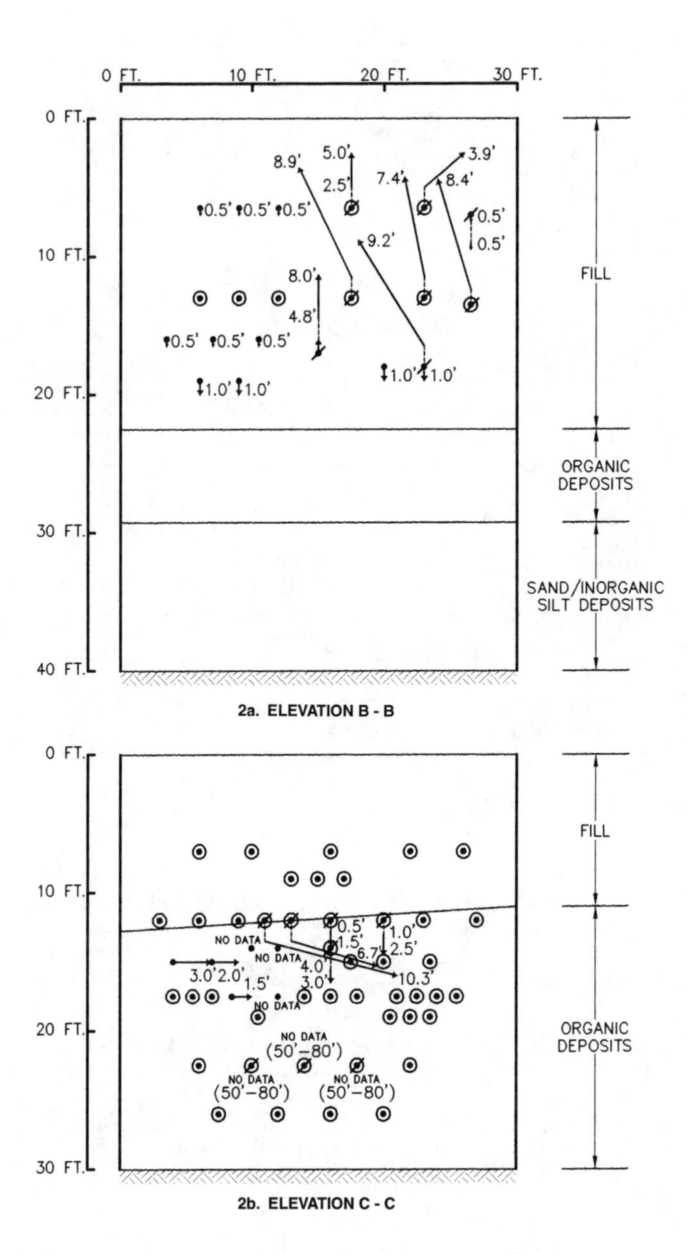

2a. ELEVATION B - B

2b. ELEVATION C - C

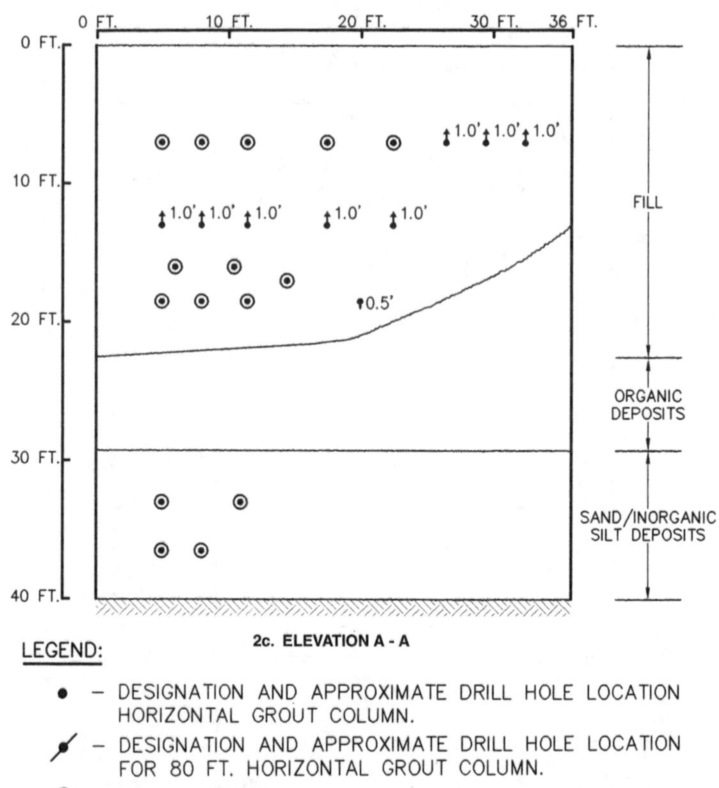

2c. ELEVATION A - A

LEGEND:

- — DESIGNATION AND APPROXIMATE DRILL HOLE LOCATION HORIZONTAL GROUT COLUMN.
- — DESIGNATION AND APPROXIMATE DRILL HOLE LOCATION FOR 80 FT. HORIZONTAL GROUT COLUMN.
- — DESIGNATION FOR HORIZONTAL GROUT ELEMENT INSTALLED WITH MEASURED HORIZONTAL / VERTICAL DEVIATION EQUAL TO 0.5 FT. +/−.
- 1.0' — DESIGNATION FOR HORIZONTAL GROUT ELEMENT INSTALLED WITH MEASURED HORIZONTAL DEVIATION EQUAL TO 1.0 FT.
- 2.0' — DESIGNATION FOR HORIZONTAL GROUT ELEMENT INSTALLED WITH MEASURED VERTICAL DEVIATION EQUAL TO 2.0 FT.
- 1.4' — DESIGNATION FOR HORIZONTAL GROUT ELEMENT INSTALLED WITH MEASURED RESULTING DEVIATION EQUAL TO 1.4 FT.
- 3.0' — DESIGNATION FOR HORIZONTAL GROUT ELEMENT WITH MEASURED VERTICAL DEVIATION EQUAL TO 3.0 FT. AT APPROXIMATELY 50−80 FT. AWAY FROM DRILL HOLE LOCATION.

Figure 2. Elevation Layout of Horizontal Grout Holes

buried obstructions would be encountered, and provide insight on obstruction impacts on drilling and grouting performance.

From a series of test borings, the soil stratification around the access excavation was determined, as shown on Figures 2a, 2b, and 2c. A number of laboratory tests were performed on soil samples from these borings, including sieve analysis for grain size of granular fill and inorganic fine sandy silt, the results of which are shown on Figures 3a, and 3b. As shown, the fill is variable with fines content (amount passing 0.075 mm (No. 200 sieve)) generally ranging from 6 to 25 percent. Silt strata generally have more than 55 percent fines, sometimes reaching 95 percent. Atterberg limit tests on organic and inorganic silt soils showed these to be largely non-plastic. The organic content of the organic silt soils typically varies from 2 to 6 percent.

PENETRATION GROUTING - GENERAL INSTALLATION PROCEDURES

All penetration grouting was done through sleeve port pipes (tube-a-manchette). The 4 cm (1.5 in.) nominal diameter sleeve port pipes were Durinval S (PVC) with a sheathed ring of 4 injection openings, 90 degrees apart, spaced every 0.35 m (1.1 ft.) along the pipe. These were usually installed using standard duplex rotary drilling with 13 cm (5 in.) O.D. casing. Attempt was made to drill with casing and cutting shoe bit only, but was found rather inefficient and generally resulted in sleeve ports requiring higher annulus grout port "cracking" pressures to fracture the thicker grout annulus. Sleeve port pipe annulus grout, with water/cement ratio of 1.5:1 and 4 percent bentonite content (by weight of cement), was effectively used for all penetration grout pipes.

Manifold (multi-line) injection with ultrafine cement-bentonite and sodium silicate was implemented using Primary (P) and Secondary (S) injection patterns listed below.

- 115 l (30 gal.) P, 115 l (30 gal.) S, at every alternating port.
- 230 l (60 gal.) P, 230 l (60 gal.) S, alternating at every second port.
- 345 l (90 gal.) P, 345 l (30 gal.) S, alternating at every second port.

Each pattern was designed to provide overall average injection volume of 115 l per 0.3 m (30 gallons per linear foot). This criteria was established based on the

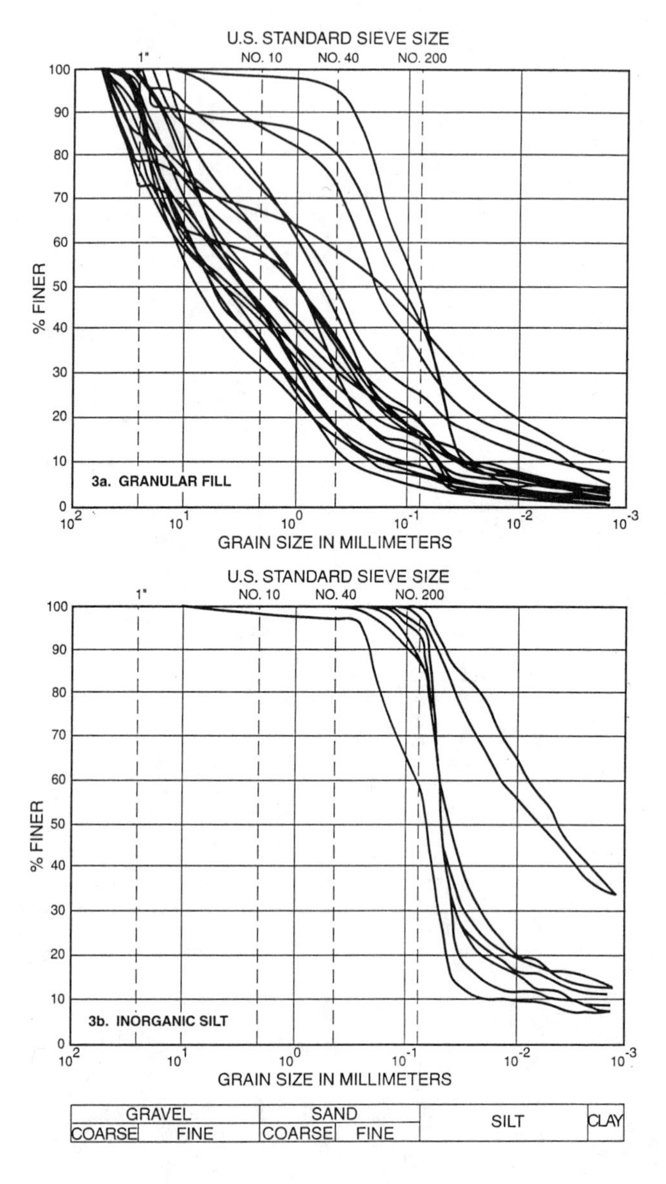

Figure 3. Grain Size Distribution Curves of Samples Taken at SSTP Site

assumption of a nominal ground porosity of 25 percent. The three patterns achieved similar grout takes.

Injection pressures usually ranged from 70 to 350 kPa (10 to 50 psi). Higher pressures (to over 700 kPA (100 psi)) usually did not increase grout "intake ratio" (ratio of actual grout intake volume to design volume). Penetration grouting appeared more effective when ports were "cracked" with water shortly prior to injection of grout.

ULTRAFINE CEMENT-BENTONITE PENETRATION GROUTING

For penetration grouting, the cementitious, suspension-type ultrafine cement-bentonite grout used ratios of components of the grout mix in the following ranges:

- Water/Cement ratio (by weight): 1:1 to 2.5:1
- Bentonite content (by weight of cement): 0% to 2%
- Admixtures/Injection Agent (by weight of cement): 1%

The ultrafine cement used in the grout mixes was Microcem B, manufactured by Lehigh Cement Co. of Allentown, Pennsylvania, with greater than 95 percent of all particles finer than 8 microns (8×10^{-6} meter) in aqueous solution, and a Blaine fineness of approximately 1,500 m^2/kg. To reduce the potential for water bleeding from the grout after injection, Pressure Grouting Agent 1 manufactured by Lehigh Cement Co., was added to all grout mixes. The ultrafine bentonite used was Bara-Kade Gold bentonite, manufactured by Bentonite Corporation of Denver, Colorado, and provided greater than 95 percent of all particles (in aqueous solution) finer than 12 microns.

Ultrafine cement-bentonite penetration grouting caused ground fractures and created lenses of grout in most locations (Photographs 1a and 1b). This is thought largely due to grout pressures exceeding overburden pressures and the soil fines content being greater than the 15 percent usually regarded as a practical limit for cement-bentonite grout permeation. Grout lenses were typically 1/4 to 1 in. thick and extended radially ½ to 3 ft. from the injection ports, decreasing in thickness and becoming quite brittle with increasing distance. In general, there was no observed preferred orientation from either vertical or horizontally grouted holes.

Permeation of ultrafine cement-bentonite grout into fill voids occurred at less than one-third of the approximately 380 attempted grout ports. Permeation appeared limited to only those zones of more clean fill (lesser fines content).

Photograph 1a and 1b - Typical Ultrafine Cement-Bentonite Lenses from Penetration Grouting

When permeation occurred, bulbs of grouted ground 0.3 to 0.6 m (1 to 2 ft.) in diameter were created; these were brittle and readily excavated.

Limited attempts at grouting ultrafine cement-bentonite into organic silt and inorganic fine sandy silt, while generally achieving desired injection volumes, only created ground fractures filled with lenses of grout.

SODIUM SILICATE PENETRATION GROUTING

Concentrations of sodium silicate used were 37, 44 and 50 percent (by total volume of grout mix) with Reactant/Hardener concentrations from 7% to 11% (by volume of sodium silicate), which produced a water soluble, stable solution having 40 minute gel time. The sodium silicate grout was not mixed with an accelerator because reduction of gel time was not required. The sodium silicate used in the grout mixes was Celtite 55-03, combined with a Terraset reactant/hardener, both manufactured by Fosroc, Inc. of Georgetown, Kentucky.

The sodium silicate grout was injected at full intake ratios in more than three-fourths of the ports grouted, using pressures 70 to 280 kPA (10 to 40 psi). This was effective in developing continuous grouted zones in granular fill, which varied slightly in shape and diameter along the length of sleeve port pipes. Sodium silicate penetration grout generally permeated fill along the full length of sleeve port pipes. However, at increasing distance from sleeve port, apparent grout effectiveness decreased (i.e. grouted ground was less firm), probably because of decreasing saturation of voids. Penetration into fill voids also varied with orientation of drill hole/sleeve port pipe. As the grout was unearthed, diameters of grout columns were measured; these are summarized for sodium silicate in Fill in Figure 4. The following discusses more aspects of the resulting sodium silicate grouted ground:

- <u>Vertical</u>: Saucer-shaped "domes" of firmly grouted granular fill were typically created which extended 0.5 to 1.4 m (1.5 to 4.5 ft.) from sleeve port pipes radially and were 0.3 to 0.5 m (1 to 1.5 ft.) thick (Photograph 2). Spread of less firmly grouted fill was typically 0.3 to 0.6 m (1 to 2 ft.) further, causing overlap of grout from adjacent ports vertically, and in limited instances laterally when 1.1 m (3.5 ft.) away. Penetration of fill voids ended abruptly at the underlying organic silt.

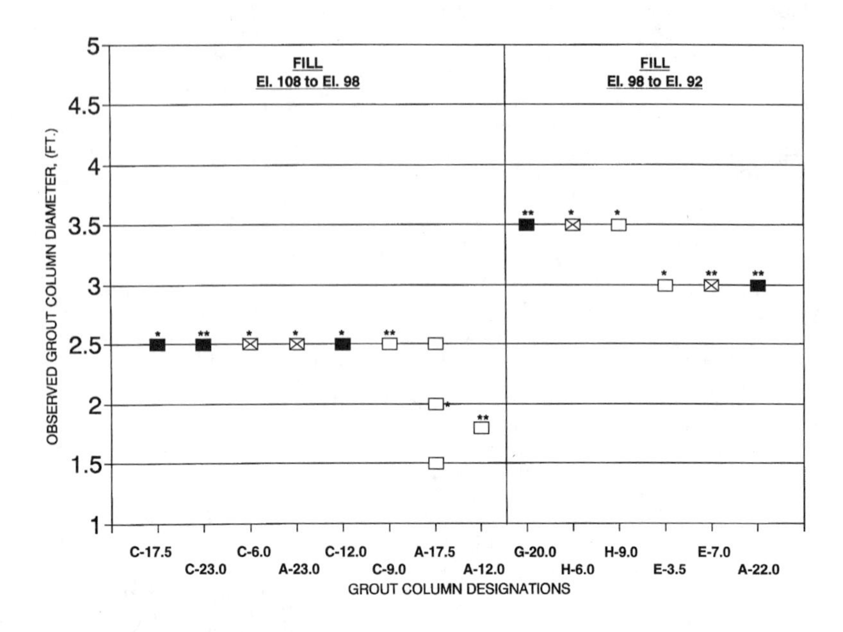

LEGEND:

OBSERVED GROUT COLUMN DIAMETERS

INDICATES MINIMUM, AVERAGE, AND MAXIMUM OBSERVED COLUMN DIAMETER, AND INJECTION SEQUENCE. (SEE BELOW FOR DESCRIPTIONS)

INJECTION SEQUENCE

☐ 30 GALLON PRIMARY AND 30 GALLON SECONDARY INJECTIONS AT EVERY ALTERNATING PORT.

☒ 60 GALLON PRIMARY AND 60 GALLON SECONDARY INJECTIONS ALTERNATING AT EVERY SECOND PORT.

■ 90 GALLON PRIMARY AND 30 GALLON SECONDARY INJECTIONS ALTERNATING AT EVERY SECOND PORT.

NOTES:

1. WHERE MINIMUM AND MAXIMUM COLUMN DIAMETERS ARE NOT REPORTED, OBSERVATIONS INDICATED APPROXIMATELY UNIFORM COLUMN DIAMETERS.
2. "**" INDICATES DESIGN GROUT MIX COMPRISED OF 37% (BY TOTAL VOLUME) SODIUM SILICATE AND 11% (BY VOLUME OF SODIUM SILICATE) REACTANT.
3. "*" INDICATES DESIGN GROUT MIX COMPRISED OF 50% (BY TOTAL VOLUME) SODIUM SILICATE AND 7% (BY VOLUME OF SODIUM SILICATE) REACTANT FOR VERTICAL GROUT HOLES, AND 44% (BY TOTAL VOLUME) SODIUM SILICATE AND 7% (BY VOLUME OF SODIUM SILICATE) REACTANT.

Figure 4. Sodium Silicate Penetration Grout Column Diameters Observed in Granular Fill

Photograph 2. Saucer-shaped Domes from Sodium Silicate Penetration Grouting in Granular Fill

o <u>Horizontal</u>: Done only in granular fill, the columns of firm grout formed were bulbous rather than cylindrical (Photograph 3) and varied in diameter from 0.5 to 1.1 m (1.5 to 3.5 ft.), being 0.2 to 0.3 m (½ to 3/4 ft.) larger at greater depths.

Bulb-shaped horizontal columns overlapped adjacent columns in four separate groups, where individual drill hole spacings were generally 1 to 1.2 m (3 to 3.5 ft.) apart, but much of the overlap was in the "less than firm" grout (Photograph 4).

Sodium silicate grout was injected into organic silt and inorganic fine sandy silt strata through vertical sleeve port pipes, but only created grout-filled lenses through ground fractures. These grout lenses generally extended radially from injection ports without direction/orientation preference.

Photograph 3. Bulbous-shaped Horizontal Columns of Sodium Silicate Grouted Granular Fill

Photograph 4. Overlapping Horizontal Columns of Sodium Silicate Grouted Granular Fill

JET GROUTING

The process of replacement or jet grouting used special horizontal drilling techniques and single rod grouting with cementitious grout to cut the in-situ soil matrix and form new strengthened soil-cement material. For the required stable colloidal grout mix, the water/cement ratio (by weight) was varied from 0.8:1 to 1.4:1. The cement used in all the mix designs was Type III, high early strength, Portland cement, conforming to ASTM C150-86 (without admixtures), and was manufactured by Dragon Products Co. of Thomaston, Maine. Bentonite was not used.

Four different sequences of drilling and jetting procedures performed in the vertical and horizontal jet grouting programs included:

Type A - Pre-drilling without casing, followed by jet-out grouting.
Type B - Pre-drilling with casing, followed by jet-out grouting.
Type C - Pre-drilling with casing, setting sacrificial casing, followed by jet-out grouting.
Type D - Jet-in grouting with casing (no pre-drilling), then setting trial reinforcement.

The jet-out grouting was performed from the bottom (or end) of the drilled hole, out toward surface (or wall entry). For the jet-in method, jet grouting was performed as the hole was advanced from ground surface (or wall entry), down toward the bottom (end) of hole. The single rod methods using jet-out (Type A and Type B) were typically performed and produced satisfactory results in the several soil units. The principal grouting variable was pressure, with 60 MPa (8,000 psi) being somewhat more effective than 30 MPa (4,000 psi), i.e., formed slightly larger grout columns. Other variables tested in apparent order of importance are: nozzle diameter (varied from 1.6 mm to 2.1 mm), water/cement ratio of mix, rod withdrawal rate (varied from 5 cm (2 in.) per 6 sec. to 5 cm (2 in.) per 22 sec.), and rod rotation rate (10 to 20 revolutions per min.). Rod withdrawal rate appeared to be second in importance in vertical jet grouting.

In the 12 vertical jet grout holes, column diameter changed considerably in different soil strata. Average measured diameters using 60 MPa (8,000 psi) grout pressure were: fill- 0.8 m (2.5 ft.), organic silt - 0.5 m (1.8 ft.), inorganic fine sandy silt - 0.7 m (2.2 ft.), and silty clay- 0.6 m (2.1 ft.). However, column diameters varied 0.2 to 0.3 m (½ to 1 ft.) smaller and larger from the averages here indicated. The vertical jet grout produced rather uniform diameters in the fill, as opposed to the saucer shaped sodium silicate penetration grout columns (Photograph 5).

Horizontal jet grouting was conducted in the fill at 12 locations, in the organic silt at 34 locations, in the inorganic fine sandy silt at 6 locations and in the silty clay at 2 locations. The grouted length generally extended 6 m (20 ft.). However, at eight locations, grouting extended 15 m to 23 m (50 ft. to 75 ft.) to determine if problems would occur with long hole alignment (Photograph 6).

Photograph 5. Comparison of Vertical Jet Grout and Sodium Silicate Grout Columns in Granular Fill

Photograph 6. Long Horizontal Jet Grout Columns

Installed diameters of horizontal jet grouting are summarized in Figures 5a, 5b, and 5c as maximum, minimum and average diameters, based on observations of the unearthed grout columns in the three soil strata. As can be seen, variation in column diameter within relatively uniform soil conditions was observed to be generally less than ± 0.2 m (± ½ ft.)

In some holes where sacrificial casing was used, slight to 0.3 m (1 ft.) reduction in grout column diameter occurred, although the jet grouting did successfully destroy the thin plastic casings (6 mm wall thickness, 100 mm diameter).

Horizontal jet grout columns overlapped to form grout "blankets" in the several test groups where drill hole spacing was from 0.4 to 0.6 m (1.5 ft. to 2.0 ft.) center-to-center (Photograph 7).

Vertical and horizontal timber crib walls and wood piles could not be effectively destroyed by jetting operations, but could be penetrated by drilling. Long horizontal drill holes were sometimes deflected considerably by the wooden obstructions. At obstructions, there was significant strength reduction in soil/cement matrix formed to either side of the timber, and little if any grout formed behind the obstructing timber (Photographs 8a and 8b).

ALIGNMENT OF HORIZONTAL GROUT HOLES

The alignment of 6 m (20 ft.) long horizontal drill holes was within 0.2 m (½ ft.) of design location in about three-quarters of the grout columns unearthed. Measured deviations from expected locations are graphically summarized in Figures 2a, 2b, and 2c.

For 6 m (20 ft.) lengths, jet grout columns were installed on-line more consistently than were the sleeve port pipes for penetration grouting. Deviations to left or right were rare. More than three-quarters of the sleeve port pipe installations went 0.2 to 0.3 m (½ to 1 ft.) lower or higher than design, depending from which side of the access excavation the drilling was conducted.

For the 15 to 23 m (50 to 80 ft.) lengths of jet grout columns and sleeve port pipes, left-right drift was considerable, frequently more than 1.5 m (5 ft.). Long (23 m (80 ft.)) sleeve port pipes tended to rise 1.8 to 2.4 m (6 to 8 ft.), whereas long (23 m (80ft.)) jet grout columns only drifted 0.3 to 0.6 m (1 to 2 ft.) lower than design elevation (in one case 1 m (3 ft.) lower). The extensive buried network of timber crib walls and piles

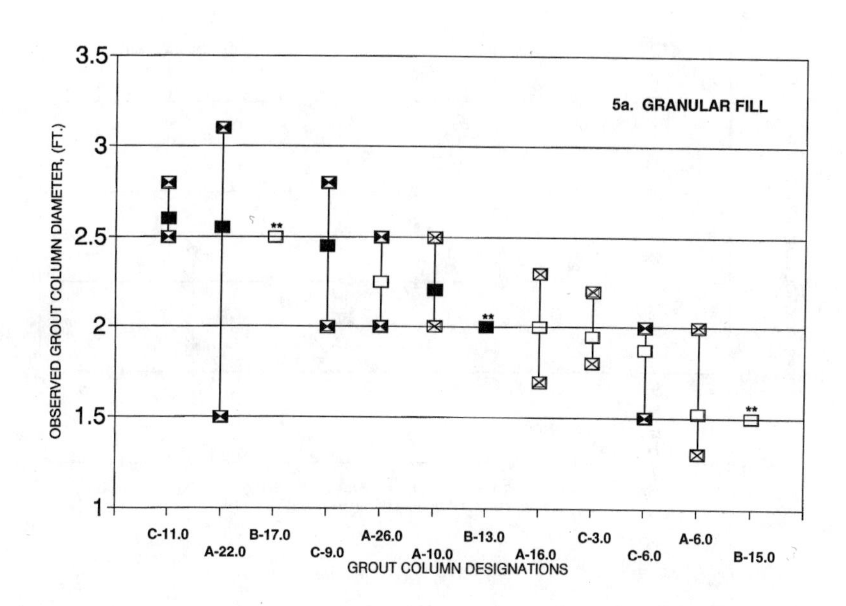
5a. GRANULAR FILL
OBSERVED GROUT COLUMN DIAMETER, (FT.)
3.5
3
2.5
2
1.5
1
C-11.0
A-22.0
B-17.0
C-9.0
A-26.0
A-10.0
B-13.0
A-16.0
C-3.0
C-6.0
A-6.0
B-15.0
GROUT COLUMN DESIGNATIONS

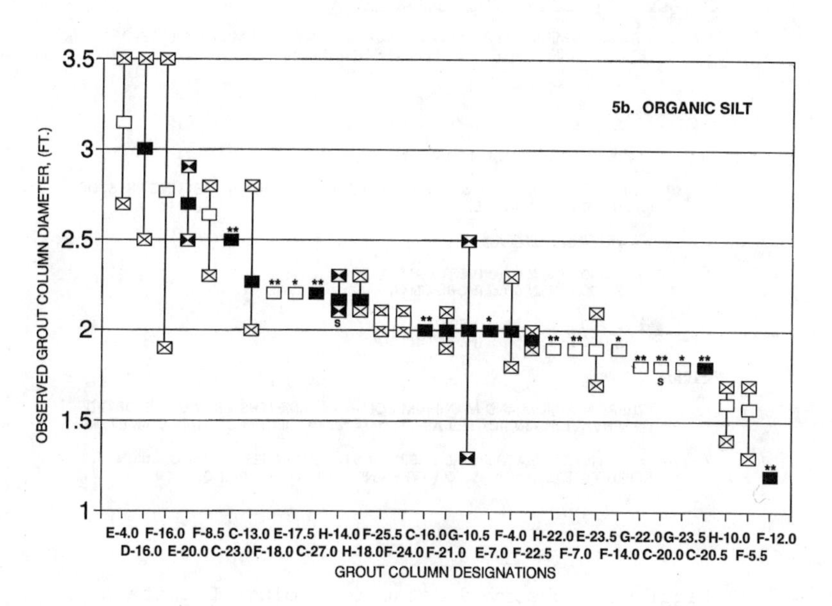
5b. ORGANIC SILT
OBSERVED GROUT COLUMN DIAMETER, (FT.)
3.5
3
2.5
2
1.5
1
E-4.0 F-16.0 F-8.5 C-13.0 E-17.5 H-14.0 F-25.5 C-16.0 G-10.5 F-4.0 H-22.0 E-23.5 G-22.0 G-23.5 H-10.0 F-12.0
D-16.0 E-20.0 C-23.0 F-18.0 C-27.0 H-18.0 F-24.0 F-21.0 E-7.0 F-22.5 F-7.0 F-14.0 C-20.0 C-20.5 F-5.5
GROUT COLUMN DESIGNATIONS

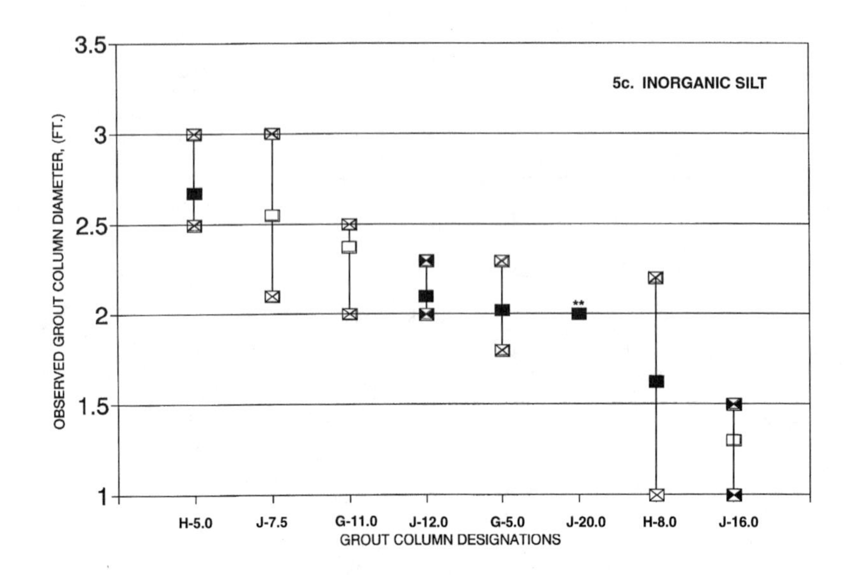

LEGEND:

OBSERVED GROUT COLUMN DIAMETERS

INDICATES MINIMUM AND MAXIMUM OBSERVED COLUMN DIAMETER, AND JETTING NOZZLE SIZE. (SEE BELOW FOR RANGES)

NOZZLE SIZE

,**	1.6 mm to 1.8 mm
,*	1.9 mm to 2.1 mm

INDICATES AVERAGE OBSERVED COLUMN DIAMETER, AND JETTING PRESSURE. (SEE BELOW FOR RANGE)

JETTING PRESSURE

3,300 to 3,700 psi (VERTICAL)
3,800 to 4,250 psi (HORIZONTAL)

7,700 to 8,200 psi (VERTICAL)
8,000 to 8,225 psi (HORIZONTAL)

NOTES:

1. WHERE MINIMUM AND MAXIMUM COLUMN DIAMETERS ARE NOT REPORTED, OBSERVATIONS INDICATED APPROXIMATELY UNIFORM COLUMN DIAMETERS.

2. "S" INDICATES SACRIFICIAL CASING INSTALLED IN JET GROUT COLUMN. REFER TO TABLE II OR IV FOR TYPE AND LENGTH OF CASING.

Figure 5. Observed Jet Grout Column Diameters

Photograph 7. Overlapping Horizontal Jet Grout Columns in Granular Fill

Photograph 8a and 8b. Horizontal Jet Grout Columns Obstructed by Timber Piles

outside the face penetrated by jet grout drilling may have caused deflections, but also provided some restraint once these were drilled through. Visual observation of horizontally drilled hole drift indicates very good alignment maintenance for most 6 m (20 ft.) lengths, but some alignment drift problems for holes more than 17 m (55 ft.) in length.

CLOSING REMARKS

The Soil Stabilization Testing Program accomplished its objectives of documenting on-site performance of penetration and jet grouting procedures that are considered likely to be used in conjunction with the tunnel jacking for the I-90/I-93 Interchange construction on Boston's Central Artery/Tunnel Project. The tunnel designer (and later the eventual contractor) will use the results of the testing program to determine specific applications for various grouting methods. Full report of the SSTP (Reference 3) will be available for a fee from the Massachusetts Highway Department, c/o Bechtel/Parsons Brinckerhoff, Central Artery/Tunnel Management Consultant, project administrator for Section Design Contract D009A.

ACKNOWLEDGMENTS

The authors wish to acknowledge the Massachusetts Highway Department, Project Owner; and Bechtel/Parsons Brinckerhoff, Project Management Consultant, for their thoughtful cooperation in preparation of this paper. The Contractors for the project were J. M. Cashman, Inc. (General) and Fonditek, International (Grouting). The assistance of specialist consultant Dr. W. H. Baker during construction of the test project is also gratefully acknowledged.

An Instrumented Jet Grouting Trial In Soft Marine Clay

C.E. Ho[1], MASCE

Abstract

An instrumented jet grouting trial using the triple-tube technique was conducted to study the induced ground displacements and build-up of soil and pore water pressures in the soft Singapore marine clay. The response of the ground shielded by a diaphragm wall as well as that directly exposed to the jet grouting works were measured and compared. The trial showed that the diaphragm wall provided an effective shield for the soft clay against the effects of jet grouting activity. Core tests indicated that column diameters of 1.8m was marginally achievable in soft marine clay, although the intersection of adjacent columns may be weak for a spacing of 1.55m. Jet grout diameter obtained in stiff dessicated clay was observed to be less than 900mm.

Introduction

The construction of a deep basement in soft marine clay close to an existing Mass Rapid Transit (MRT) viaduct and station in Singapore necessitated the use of a jet grout slab strut to limit the diaphragm wall deflections. The basement excavation was between 12 to 17m deep and was to be braced with 4 and 5 levels of temporary steel struts. The jet grout slab strut was between 3 to 4m thick across the site. For the zone up to 10m in front of the diaphragm wall facing the MRT structures, the jet grout was installed to 9m thick. Induced ground displacements and pressures were required to be limited to within very tight tolerances specified by the Mass Rapid Transit Corporation.

Jet Grouting Techniques

A comprehensive discussion on the applications of jet grouting techniques in various soil types has been given by Bell (1993). The most common method used for jet grouting works in Singapore is the double-tube technique. This technique involves the initial drilling of a borehole about 100mm to 150mm in diameter to the toe level of the intended jet grout column. Simultaneous cutting of soil and grouting is then

[1] Associate, Arup Geotechnics, Ove Arup and Partners Singapore, 5001 Beach Road #04-03 Golden Mile Complex, Singapore 0719, Singapore.

executed by producing a jetted stream of grout at high pressure. The effectiveness of the cutting process is enhanced by an envelope of compressed air which assists in the opening up of crevices formed by the jetted grout. The distance of the cutting action is dependent on the pressure of grout and air injected.

A common difficulty with using the double-tube technique in soft marine clay in Singapore is the high viscosity of the cohesive grout-marine clay mixture that is produced. This is due to the high clay content of the Singapore marine clays. The consequence of such a heavy and viscous material is to create potential blockage to the backflow of slurry up the borehole, thus causing a rapid build-up of pent up pressures in the surrounding soils due to the undrained state of the clay. Even when a casing is used to prevent the borehole from collapsing inwards, blockage could still develop within the restricted space unless regular flushing with water is carried out. The process of the double-tube jet grouting works in soft marine clay is almost always accompanied by significant heave of the ground, (Berry et al, 1987)

The use of the double-tube technique has in the past been successful in treating soft marine clays for several projects in Singapore where ground movements were not critical, such as for stabilization of wide canals, (Liang et al, 1993 and Pagliacci et al, 1994). However, in the case of one instance where double-tube jet grouting technique was used to form jet grout slabs between diaphragm walls for deep basement excavation in soft marine clay, significant displacement of the existing underground Mass Rapid Transit tunnels were experienced some 22m away from the diaphragm wall due to build-up of soil pressures. Because of such experience with the double-tube technique in soft marine clays, it is generally inadvisable for applications where sensitive structures are in close proximity.

For the current project, the triple-tube jet grouting technique was adopted in lieu of the conventional double-tube technique. The triple-tube technique involves the drilling of a borehole similar to the double-tube technique to the required toe level of the grout column. The jet grouting process however is performed using two levels of nozzles. The upper and lower nozzles maybe separated by a distance of between 300 to 750mm. The cutting of the soils is implemented with the upper nozzles by utilizing a jetted stream of water under high pressure instead of grout. An envelope of compressed air is similarly injected to assist in projecting the water jet to a greater distance. Grouting is via the lower nozzle under pressures much lower than the double-tube technique. This procedure tends to produce a more fluid cutting as the water content of the slurry is increased significantly. Viscosity is therefore reduced and potential blockage is prevented. The separation of the grouting process from the cutting process enables a denser grout to be placed within the pre-excavated column beneath the upper nozzles which therefore limits the mixing of the grout with the soft clay. Grout loss in the return slurry is also very much reduced compared with the double-tube technique. The objective therefore is to ensure a more efficient return of the slurry up the borehole so that blockage is avoided and build-up of pressure in the clay is prevented.

Jet Grouting Trial

An instrumented jet grouting trial using the triple-tube technique was conducted to study the induced ground displacements and build-up of soil and pore water pressures in the soft marine clay. The typical profile of the soil to be treated is shown in Fig.1. Properties of soft clays encountered at the site are given in Table 1. A length of 0.8m thick diaphragm wall about 18.6m long was first constructed from one 6.4m and two 6.1m long panels. A jet grout block 6.45m by 6.75m in plan and 9m thick was then formed at the middle of the diaphragm wall from +93 to +84 mOD. A total of 18 numbers of jet grout columns were executed to form a grout block. The individual jet grout columns were about 1.8m in diameter and were installed in triangular grids of 1.55m spacing, in an alternate fashion such that no adjacent columns were installed within a period of less than 24 hours to avoid weakening of the preformed columns. In general , two columns were installed within a day, one in the morning and one in the afternoon. Each jet grout column took between 2 to 3 hours to complete. Only one rig was used. The spacing between the upper and lower nozzles was 750mm. Cutting by

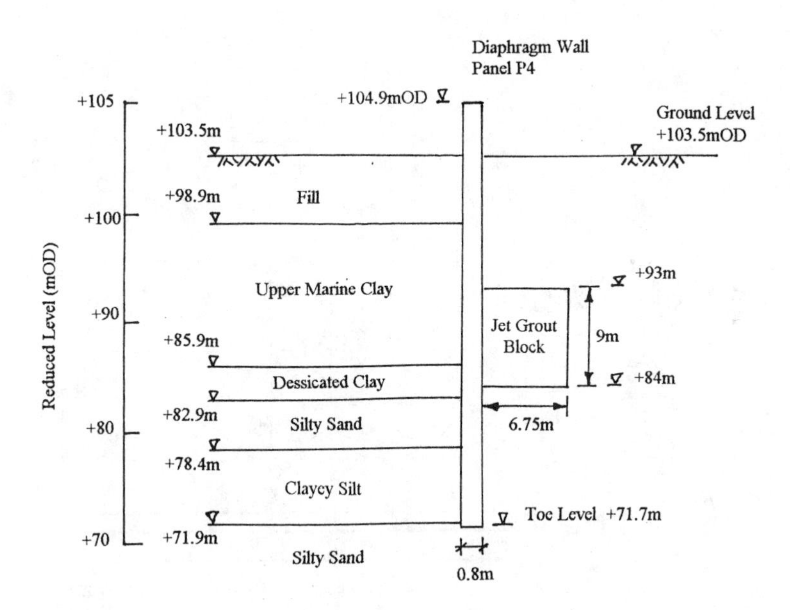

Fig.1 Soil profile at jet grout trial

TABLE 1. Typical Properties of Soft Marine Clay

Borehole No.	Sample No.	Depth (m)	Bulk Density (kN/m3)	Moisture Content (%)	Liquid Limit (%)	Plastic Limit (%)	Plasticity Index (%)	Undrained Shear Strength (kPa)	Grading (%)				Soil Type
									Gravel	Sand	Silt	Clay	
B3	UD1	7.5 - 8.20	15.43	58									UMC
	UD2	9.0 - 9.8			93	40	53	8	0	1	48	51	UMC
	UD3	10.5 - 11.3			94	40	54						UMC
	UD4	13.5 - 14.3	14.77	44									UMC
B7	UD1	7.5 - 8.3			96	40	56		0	1	39	60	UMC
	UD2	10.5 - 11.3	15.90	42	91	37	54						UMC
	UD3	13.5 - 14.3			92	39	53	6					UMC
B9	UD1	9.0 - 9.8	15.35	46	91	39	52						UMC
	UD2	12.0 -12.8	15.38	66	95	41	54	8	0	1	46	53	UMC
	UD3	24.0 - 24.8	16.01	41	85	34	51						LMC
B25	UDP1	9.0 - 9.8											UMC
	UDP2	13.5 - 14.3	14.94	62	96	40	56	14	0	1	40	59	UMC
	UDP3	16.5 - 17.3	18.65	31	60	22	38		0	14	36	50	DC
B35	UDP1	9.0 - 9.8	15.87	68	86	36	50		0	1	61	38	UMC
	UDP2	12.0 - 12.8			102	39	63	9					UMC
B38	UDP2	16.5 - 17.3	15.33	63	94	37	57	10	0	1	44	55	UMC
B39	UDP1	12.0 - 12.8	15.34	67	85	37	48	8	0	2	50	48	UMC
B44	UD1	10.5 - 11.3	15.28	69	88	35	53	9	0	1	54	45	UMC
	UD3	24.0 - 24.8	16.73	53	77	31	46		0	1	47	52	LMC

UMC - Upper Marine Clay LMC - Lower Marine Clay DC - Dessicated Clay

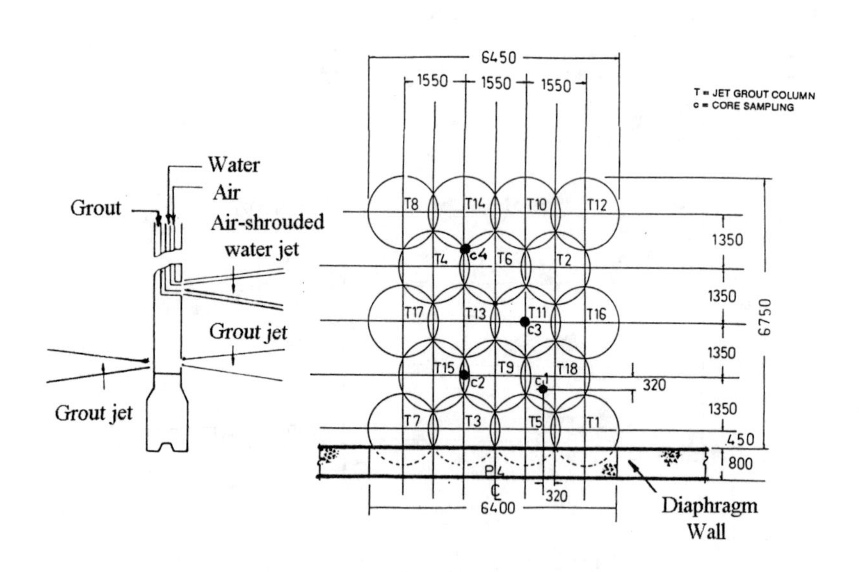

Fig.2 Layout and sequence of jet grout columns

water and air was carried out with a single jet in the upper level, whilst grouting was executed with two jets located diametrically across in the lower nozzle (Fig.2). The nozzle sizes were 10mm for air, 2.2mm for water and 4.5mm for grout. Measured flow rates were approximately 80 l/min (water), 110 l/min (grout) and 10500 l/min (air). Fig 2 shows the layout and sequence of execution of the jet grout columns. The sequence of grouting is numbered T1 to T18. Initially, it had been planned to execute the jet grouting within 150mm diameter predrilled PVC tubes placed to the top of the jet grout column. The PVC tube was to act as a casing to prevent borehole collapse in the soft clay and maintain a clear passage for the return slurry. This procedure was however terminated after T4 due to frequent blockage. Predrilling was subsequently done directly with a 200mm diameter drill bit without any casing. The operating parameters adopted for the trial are tabulated in Table 2. Water/cement ratio ranged between 0.73 to 0.87. Monitored cement density was between 14.9 and 15.7 kN/m^3 with an average value of 15.2 kN/m^3. Densities of the sludge return ranged between 12.7 to 16.1 kN/m^3.

Geotechnical instruments were installed in the ground as well as in the diaphragm wall to monitor their response, Fig.3. Inclinometers, pneumatic piezometers, water standpipes and earth pressure cells were installed in the soft clay at several distances up to 20m from the diaphragm wall. The instruments were located on both sides of the diaphragm wall. The response of the ground which was shielded by the wall, as well as that which was directly influenced by the jet grouting works were measured and compared. Initial datum readings were obtained prior to any jet grouting activity. During the trial, one set of readings was taken after completion of each grout column. The period of monitoring was extended beyond the completion of the jet grout installation to assess the rate of dissipation of the induced ground displacements and soil and pore water pressures in the soft clay.

TABLE 2. Operating Parameters for Jet Grout Trial

Jet Grout Column	Date Installed	Time Installed	Water Pressure (bar)	Air Pressure (bar)	Grout Pressure (bar)	Withdrawal Rate (mins/m)	Rotational Speed (rpm)
T1	5/7/94	0930 to 1120	450	10	100	8	9
T2	5/7/94	1405 to 1610	450	10	100	8	9
T3	6/7/94	1315 to 1515	380 to 400	6 to 7	80	9	9
T4	6/7/94	1635 to 1930	450	10	100	9	9
T5	7/7/94	1035 to 1315 1845 to 2030	380 to 400	6 to 7	60 to 80	9	9
T6	7/7/94	1530 to 1815	450	10	80 to 100	9	9
T7	8/7/94	1030 to 1250	400	6.5	80 to 100	9	9
T8	8/7/94	1515 to 1715	430	6.5	80	9	9
T9	9/7/94	0845 to 1030	460	6.5	80	9	9
T10	9/7/94	1405 to 1730	450	6.5	70	8	9
T11	10/7/94	0830 to 1110	450	6.5	65	9	9
T12	10/7/94	1310 to 1630	450	6.5	60	8	9
T13	11/7/94	0845 to 1125	450	6.46	80	9	9
T14	11/7/94	1350 to 1640	450	6.5	80	9	9
T15	12/7/94	0810 to 1025	450	6.5	50	9	9
T16	12/7/94	1310 to 1512	450	6.5	60	9	9
T17	13/7/94	0810 to 1020	450	6.5	60 to 70	8	9
T18	13/7/94	1300 to 1520	450	6.5	60 to 70	8	9

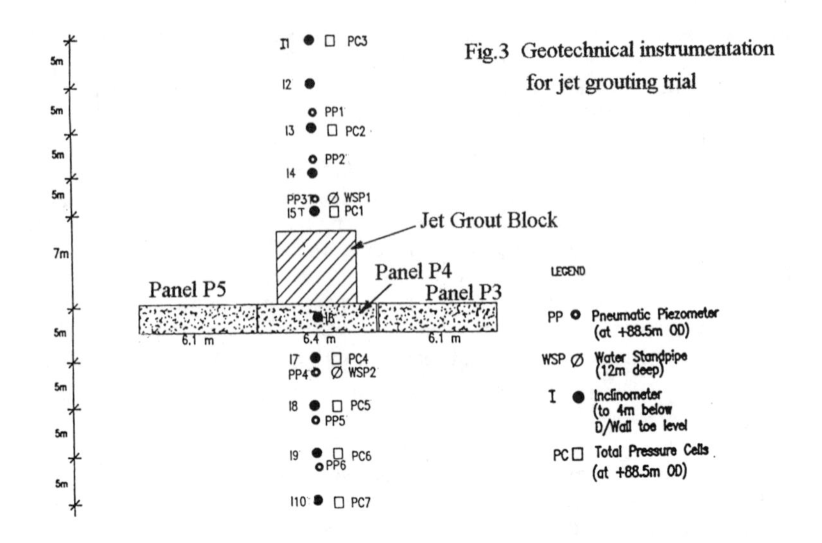

Fig.3 Geotechnical instrumentation for jet grouting trial

Diaphragm Wall and Ground Displacements

Fig. 4 shows the magnitude of lateral diaphragm wall and ground displacements associated with the execution of the jet grout columns in the direction perpendicular to the diaphragm wall. Inclinometers I1 to I5 were installed in the soft marine clay next to the jet grouting works. The displacement pattern of the closest inclinometer I5, located at 0.5m from the edge of the jet grout column was very erratic indicating significant disturbance of the soft clay. The largest displacements were of the order of 62mm in the transverse direction. The inclinometer did not survive beyond execution of T4. The adjacent inclinometer I4 recorded a maximum displacement of 34mm at about 8m below ground after execution of T3. The displacements stabilized between 21mm and 28mm beyond completion of T6 at the same level. For the furthest inclinometer I1 at 20m away from the jet grouting works, the maximum recorded displacement was only 7.5mm, occurring at the ground surface, after the execution of T15. I2 located 15m away and I3 located 10m away registered maximum displacements of 17mm after T15 and 22mm after T4 respectively, with both displacements also occurring at the ground surface. The rebound of the ground 5 days after completion of T18 was less than 2mm for I1, I2 and I3.

The effect of jet grouting operation on the ground shielded behind the diaphragm wall is indicated by inclinometers I7 to I10. I7 was located 5.2m from the back of the diaphragm wall. The positions of I8, I9 and I10 were 9.9m, 14.4m and 16.8m from the back of the diaphragm wall respectively. Inclinometer I6 was installed in the diaphragm wall. The nearest inclinometer I7 recorded a maximum displacement of 7.5mm at a depth of 8m after T14. For I8, a maximum displacement of 5mm occurred at about 7.5m depth after T8. The displacement of I9 perpendicular to the wall was less than 4mm, occurring near the ground surface. The furthest inclinometer I10 however recorded about 6mm in the opposite direction at 8m depth after T15. The transverse displacements of I10 were more significant with a maximum displacement of 6mm at about 5m depth, registered 2 days after the completion of T18. Rebound of the ground after completion of T18 was generally less than 3mm. The diaphragm wall therefore acted as a very effective shield from the effects of the grouting works to reduce displacement in the surrounding grounds. Nevertheless, the residual displacements induced in the soft clay after completion of jet grouting activity was significant and were irrecoverable. The deflections of the diaphragm wall are shown in Fig.5. The maximum registered deflection of 10mm perpendicular to the wall occurred after execution of T16 at a depth of 9 to 10m, which is at about 2 to 3m above the top level of the grouted zone. The wall deflections in the transverse direction was less than 3mm and were therefore insignificant. Fig.6 summarizes the maximum deflections in the direction perpendicular to the diaphragm wall.

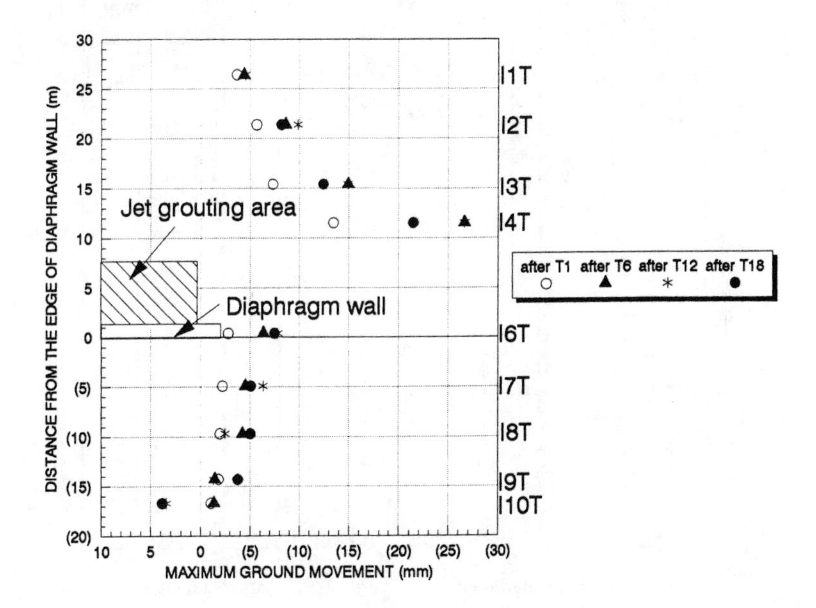

Fig.4 Maximum lateral ground displacements perpendicular to diaphragm wall

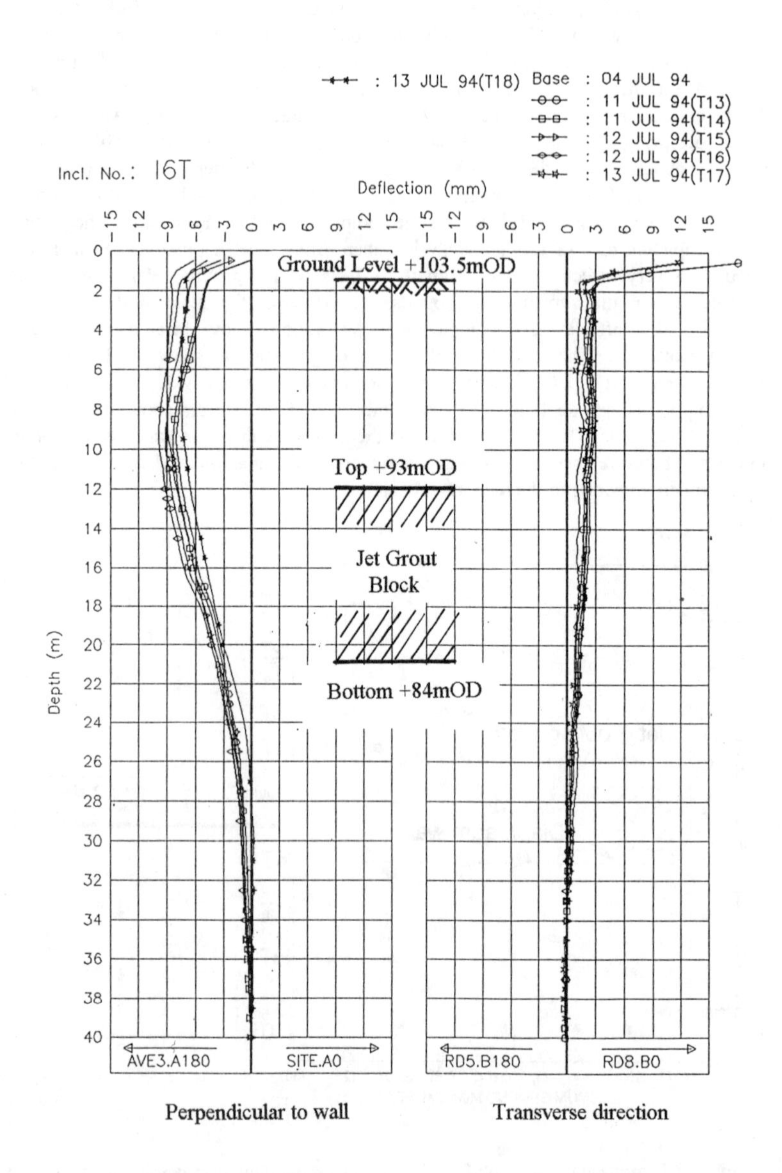

Fig.5 Lateral diaphragm wall deflections at inclinometer I6

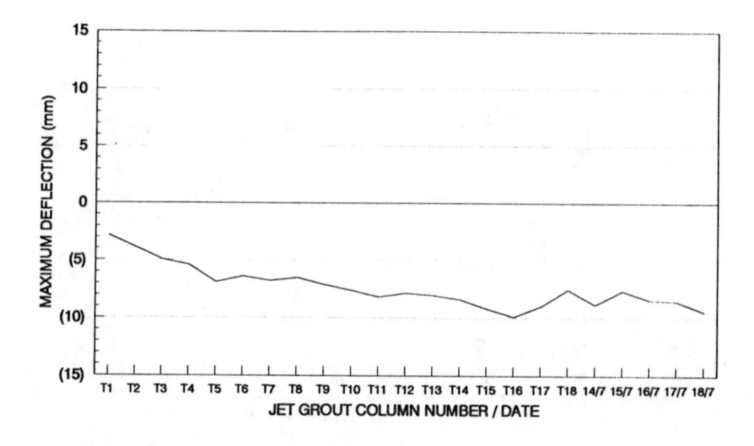

Fig.6 Maximum diaphragm wall deflections (perpendicular to wall) with jet grouting activity

Soil Pressures

The soil pressures generated in the soft clay by the jet grouting works were recorded by earth pressure cells located at a level corresponding to the middle of the grout block. Fig.7 depicts the pattern of pressure build up at different distances from the jet grouting works on both sides of the diaphragm wall. The maximum increase in soil pressure of 0.73 bars was recorded at PC1 located within 0.3m of the jet grout block. The increase in soil pressures appear to be related to execution of the first few columns where air pressures between 6.5 to 10 bars were adopted. No appreciable change in magnitude was recorded for subsequent grout columns once the air pressures were maintained at 6.5 bars. The increase of soil pressure may also be attributed to the consequence arising from the initial procedure of jet grouting within the 150mm diameter pvc tubes, which restricted the passage of return slurry to within the size of the tube and led to frequent blockage. The pvc tubes were abandoned after T4 and predrilling was carried out through the soil supported only by slurry without the use of casing. It was obvious from Fig.8 that the excess pressure was reduced from T5 onwards. In general, the other earth pressure cells did not register increase of larger than 0.22 bars in the marine clay during jet grouting on both sides of the diaphragm wall. The maximum difference between the peak pressure induced during jet grouting and the residual ambient pressure was less than 0.16 bars generally, except for the closest pressure cells PC3 and PC4 which registered 0.5 and 0.41 respectively.

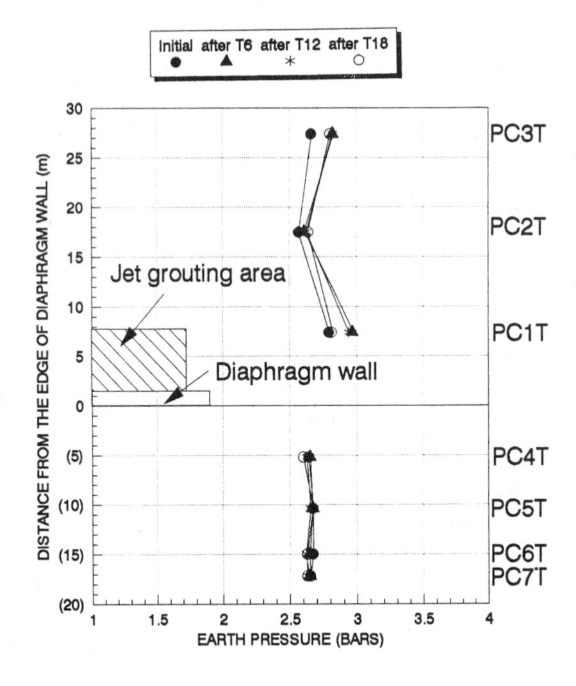

Fig.7 Soil pressures in soft marine clay

Pore Water Pressures

Fig.9 depicts the pore water pressures in the soft clay as recorded by pneumatic piezometers installed at the level of the middle of the jet grout block. All pneumatic piezometers in the soil directly exposed to jet grouting indicated significant increase in piezometric pressures. Fig.10(a) indicates the maximum increase of 69 kPa was registered after the execution of T4 in PP3 located within 1m of the jet grout block. Excess pore water pressures of 13 kPa (in PP2) and 8 kPa (in PP1) were indicated, at 4.8m and 9.5m further behind PP3 respectively. The dissipation of the excess pore pressures in the soft clay was very rapid after completion of T18. The highest average rate was recorded in PP3 with a value of 20 kPa over 5 days (or 4.0 kPa/day). PP2 and PP1 registered dissipation of 13 kPa and 6 kPa over 5 days (or 2.6 kPa/day and 1.2 kPa/day) respectively.

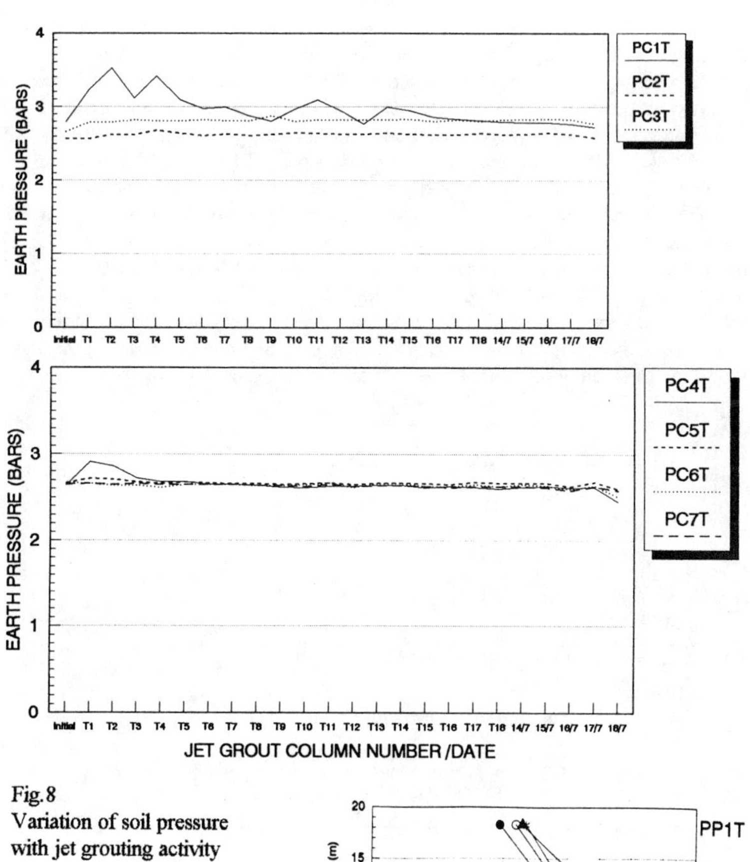

Fig.8
Variation of soil pressure with jet grouting activity

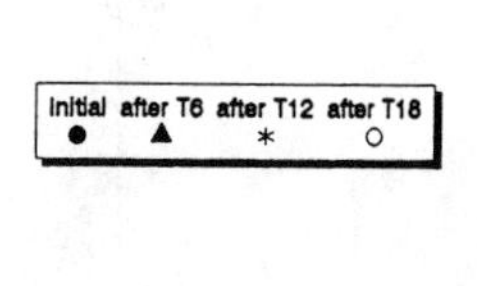

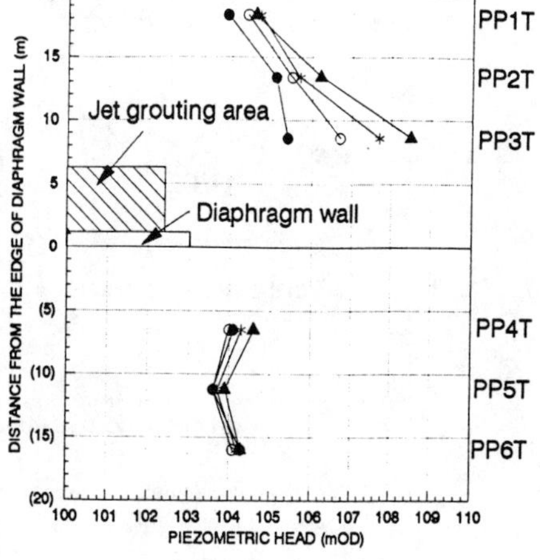

Fig.9
Piezometric pressures in soft marine clay

A similar trend existed for pore pressure response in soft clays shielded by the diaphragm wall although the magnitudes were significantly reduced, Fig.10(b). PP4, PP5 and PP6 located 6.2m, 10.9m and 15.5m behind the diaphragm wall recorded excess pore pressures of 19 kPa, 10 kPa and 4 kPa above their initial values. The average rates of excess pore pressure dissipation for the same piezometers were respectively 21 kPa, 7 kPa and 6 kPa over 5 days (or 4.2 kPa/day, 14 kPa/day and 1.2 kPa/day). There was no obvious difference in the rates of dissipation in the soft clays on either side of the diaphragm wall. Fluctuation in water table monitored in 12m deep water standpipes during the period of jet grouting activity was less than 1m. The maximum increase in raised water table was 0.8m and 0.22m in WSP1 and WSP2 respectively. WSP1 was installed about 1.2m behind the jet grout block whereas WSP2 was installed about 5.6m behind the diaphragm wall. The residual increase in the water table 5 days after completion of T18 was 0.57m in WSP1 and 0.43m in WSP2. Fig 11 shows the variation of the water levels with time and jet grouting activity.

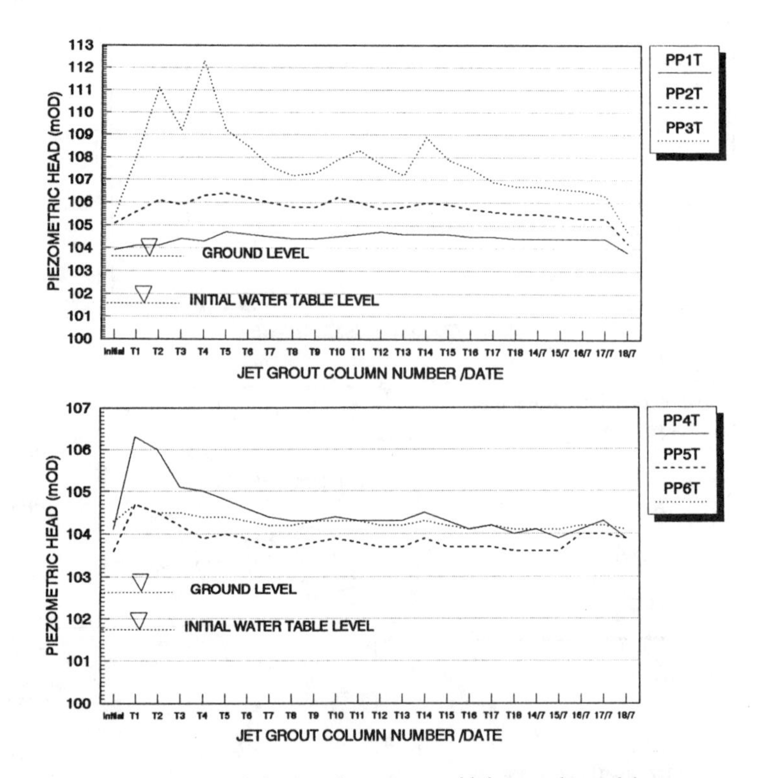

Fig.10 Variation of piezometric pressures with jet grouting activity

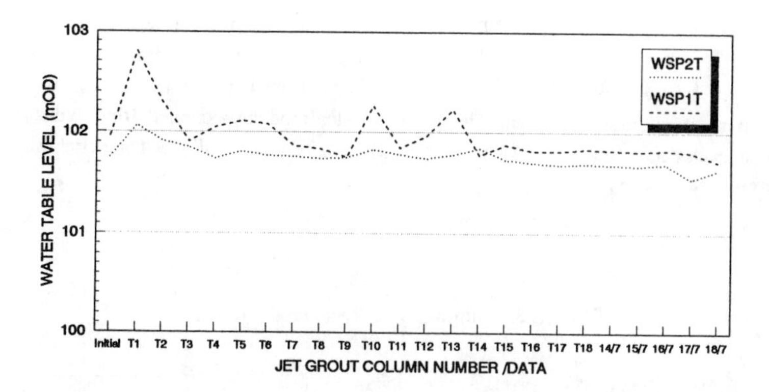

Fig.11 Variation of water table with jet grouting activity

Dimensions and Properties of Jet Grout Columns

Fig.2 shows the four locations, C1 to C4 where continuous cores were taken to verify the dimensions and quality of the jet grout columns formed. A summary description of the cores recovered is given in Table 3. Except for the initial core run, poor recovery was obtained for C3 located in the middle of column T11, probably due to poor drilling technique. A core recovery of 96% was obtained in the first 1.5m core run with consistent good quality grout. Compressive strength of 3.843 MPa and grout stiffness of 295 MPa was obtained from tested grout samples.

It was observed from C1, C2 and C4 that consistently no grout was present in the cores at levels below +86 to +87 mOD, where stiff dessicated clay was encountered. It may be inferred from C1, which was drilled 452mm away from the theoretical centre of T18, that the diameter achieved in the dessicated clay could be limited to less than 900mm. Alternatively, it may also be deduced that the verticality of the grout columns formed was of the order of 1:40 with respect to a platform level of +104.435mOD, implying that the spacing of 1.55m was inadequate to ensure the necessary contact at column intersections. The grout cores in C2 located at a double intersection between T9 and T15, indicated reasonably good core recovery. However, the grout cores obtained from C4 located at a triple intersection between T4, T6 and T14 were weak and highly fractured. This means that in the soft marine clay, good quality jet grout columns with diameters of 1.8m may not be consistently obtained and continuity may be poor at the column contacts.

Table 4 summarises the properties of the core samples of grout tested. Densities of grout cores ranged between 14.24 to 16.5 kN/m^3. Unconfined compression tests gave q_u = 1.26 to 15.06 MPa and E = 89.9 to 422 MPa. Insitu pressuremeter tests were also performed for comparison. The pressuremeter modulus obtained from unload-reload cycles gave E_p = 175.85 to 518.71 MPa. The quality of grout material was therefore satisfactory.

TABLE 3. Summary of Recovered Cores

Coring Point	Jet grout column	Core No.	Core Run (m)	Total Core Recovery (%)	Length of Grout (m)	Length of Stiff Dessicated Clay (m)	Remarks
C1	T18	1	1.5	95.3	1.42		Poor core recovery for Core No.4
		2	1.5	100.0	1.5		
		3	1.5	66.7	1.0		
		4	1.5	3.0	0.045		
		5	1.5	30.0		0.45	
		6	1.5	100.0		1.5	
C2	T15/T19	1	1.5	93.3	1.4		
		2	1.5	63.0	0.95		
		3	1.5	90.0	1.35		
		4	1.5	83.3	1.25		
		5	1.5	82.0	0.43	0.8	
		6	1.5	76.4		1.15	
C3	T11	1	1.5	96.0	1.44		No core recovery for Core No.2 to 6
		2	1.5	0			
		3	1.5	0			
		4	1.5	0			
		5	1.5	0			
		6	1.5	0			
C4	T4/T6/T14	1	1.5	66.7	1.0		Core material highly fractured, very weak.
		2	1.5	83.3	1.25		
		3	1.5	100.0	1.5		
		4	1.5	46.7	0.7		
		5	1.5	93.0	0.28	1.12	
		6	1.5	67.0	0.075	0.93	

TABLE 4. Properties of Jet Grout Core Samples

Coring Point	Jet Grout Column	Reduced Level (mOD)	Bulk Density γ_b (kN/m3)	Unconfined Compressive Strength q_u (MPa)	Laboratory Young's Modulus E (MPa)	Pressuremeter Modulus E_p (MPa)
C1	T18	92.513				175.851
		90.013 to 88.513	14.91	2.879	121	
			14.26	1.542	89.9	
		89.313				518.717
C2	T15/T19	92.30 to 91.90	15.64	9.076	357	
		91.30 to 90.40	16.10	13.355	422	
		90.1 to 89.1	14.24	1.260		
		87.1 to 86.1	14.52	15.063		
C3	T11	86.902 to 85.402	14.68	3.843	295	
C4	T4/T6/T14	90.0 to 88.50	16.50	3.837	291	
		89.7				226.320

Conclusion

a) The effects of jet grouting activity can be significantly reduced by the shielding effect of a diaphragm wall. Displacements in soft marine clay directly exposed to jet grouting works were limited to less than 7.5mm at a distance of 20m from the jet grouting activity. The same magnitude of displacement was observed at about 5.2m behind the diaphragm wall.

b) Satisfactory jetting performance can be obtained for air pressures up to 6.5 bars without causing undue disturbance to the soft clay.

c) Rebound of the soft marine clay after completion of jet grouting was only 2mm to 3mm, with resulting residual displacements which were irrecoverable.

d) Earth pressure cell measurements appear to be insensitive to the jet grouting effects with values of less than 0.22 bars being recorded generally.

e) The rates of dissipation of excess pore pressure in the marine clay after jet grouting have ceased were approximately similar, whether the soft clays were directly exposed to or shielded from the jet grouting activity by the diaphragm wall. Water table rise during jet grouting works was less than 0.8m.

f) Formation of 1.8m diameter columns in soft marine clay was marginally achievable, although the intersection of adjacent columns may be weak and potential difficulty in attaining continuity at the column contacts may be envisaged at a spacing of 1.55m. Jet grout column diameters may be limited to less than 900mm in stiff dessicated clay. The quality of jet grout material obtained were satisfactory.

References

Bell A.L. (1993). "Jet Grouting", In Ground Improvement, (Ed. Moseley M.P.). Publ. Chapman and Hall, Glasgow, pp 149-174.

Berry G.L., Shirlaw J.N., Hayata K. and Tan S.H. (1987). "A Review Of Grouting Techniques Utilized For Bored Tunneling With Emphasis On The Jet Grouting Method". Proc. of the Singapore Mass Rapid Transit Conference, Singapore, pp 207- 214.

Liang K.M., Ganeshan V. and Oei R.T.B. (1993). "Soil Improvement By Jet Grouting For Reconstruction Of An Open Channel In Singapore Marine Clay". Proc. 11th Southeast Asian Geotechnical Conference, Singapore, pp 381-386.

Pagliacci F., Trevisani S. and Chong L. (1994). "Recent Development In Jet Grouting Techniques - Singapore Pulau Seraya Power Station, A Case History". Proc. 3rd Int. Conf. On Deep Foundation Practice, Singapore, pp 219-225.

Ataturk Dam - Hydrogeological and Hydrochemical Monitoring of Grout Curtain in Karstic Rock

By Wynfrith Riemer[1], Michel Gavard[2], Mümtas Turfan[3]

Abstract

For the Ataturk dam, a grout curtain of an area of 1.2 km^2 was constructed. Complex karst problems rendered the construction of the curtain and the quality assurance difficult. Methods of performance monitoring had to be developed which relied primarily on piezometric observations. These were complemented by tracer tests and water quality monitoring. All this allowed quantified assessment of the performance of the curtain and provided guidance for the construction work.

1. Introduction

The Ataturk dam, on the Euphrates (Firat) river in Turkey, with a fill volume of 84 million m^3, impounds a reservoir of 49 billion m^3. The project serves for hydropower (installed capacity 2400 MW) and irrigation (see Basmaci, 1991, for description of the project). In the latter context it forms the key component of the Southeast Anatolian Development Project. The foundation conditions for the dam are difficult. It rests on karstic limestone and dolomite, extending to a depth of about 600 m below the dam foundation level. The main grout curtain, with a length of 5.5 km, reaches 175 to 300 m deep into the valley floor and still ends high above the base level of the karst.

Concurrent with the construction of the grout curtain, a hydrogeological monitoring system was established. Regular hydrogeological and hydrochemical monitoring had accompanied the construction of the curtain and was intensified with the commencement of reservoir filling in November 1989. By 1994 the lake level had risen 153 m, 7 m short of full pond.

[1] Geologist, DSc, M.ASCE, L-8529 Ehner, Luxemburg
[2] Vice President, Electro-Watt Engineering, CH-8034 Zürich, Switzerland
[3] Deputy General Manager, State Hydraulic Works (DSI), Ankara, Turkey

2. Geological Setting

Lithology

The sequence starts with Cretaceous, massive dolomitic limestones which are subdivided by intercalations of chert and bituminous marl. A unit of about 50 m bituminous marl and chert overlies the dolomitic limestones. This is succeeded by some 150 m of stratified, platy marly limestone with chert bands which constitutes most part of the dam foundation. At the curtain, the sequence ends with massive marly limestones. Tertiary claystones and marls form the impervious reservoir rim.

Structure

The sequence contains several unconformities. A regional fault, the Bozova Fault, runs across the left end of the curtain. The Cretaceous sediments are faulted into an anticline which obliquely crosses the curtain. Intensive faulting and shearing are superimposed on the anticline (Figure 1). Tectonic fracturing, combined with dissolution of carbonates weakened the marly limestones to render them partially erodible.

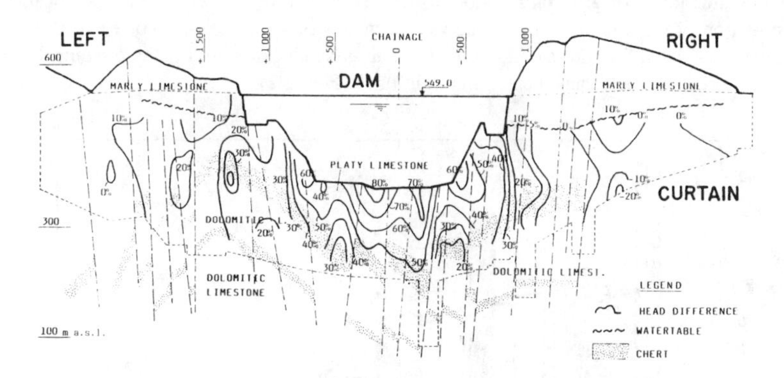

Figure 1: Geological section along grout curtain. Contours show head difference across the curtain in % of total reservoir-tailwater head difference

Hydrogeology

Prior to impounding a regional karst aquifer had discharged about 1 m^3/s to the river at the dam site, according to observations made during excavation of the diversion tunnels and of the dam foundation. The karst aquifer was subdivided into a semi-confined level in the lower part of the dolomitic limestone and an upper phreatic level in dolomitic and platy limestone, with a layer of bituminous marl acting as aquitard. Lenses of perched water existed in the marly limestone.

3. Construction of the Grout Curtain

With an area of 1.2 km², the grout curtain ranks among the largest ever constructed. Its construction proved more difficult than anticipated. In the vicinity of the major faults and particularly in the contact with cherty limestone, large open voids had developed which absorbed massive volumes of grout (frequently in excess of 2 tonnes/m) before any effect was achieved. On the other hand, the grout penetration in the weak marly limestone was very poor. One area even required soil grouting techniques with sleeve port tubes. The contractor was not prepared to adjust the composition of the slurries to match the wide variation in the rock groutability, and this circumstance further complicated the work. Most part of the grouting work was done with cement-bentonite slurries (mainly water/cement = 1, with 2% bentonite), during later stages high density cement slurries (water/cement = 0.65) with 0.25% liquefier and in one specific area also high-penetration Microsol grout (trademark of Soletanche) were used. Grouting pressures ranged from 3 bar near foundation level to 40 bar at depth. The overall grout take in the main curtain cumulated to 113,000 tonnes or an average of 95 kg/m². With an average spacing of boreholes in the main curtain of 0.95 m, the take per linear meter arrived at 90 kg/m which falls into the "moderate" range of grout absorption, as defined by Deere (1982). As the statistical evaluation of specific grout takes shows, the modes of grout take do not exceed 50 kg/m² (calculated for panels of 25x36m²) and differ significantly for the three segments of the curtain, i.e. left flank, center and right flank (Figure 2).

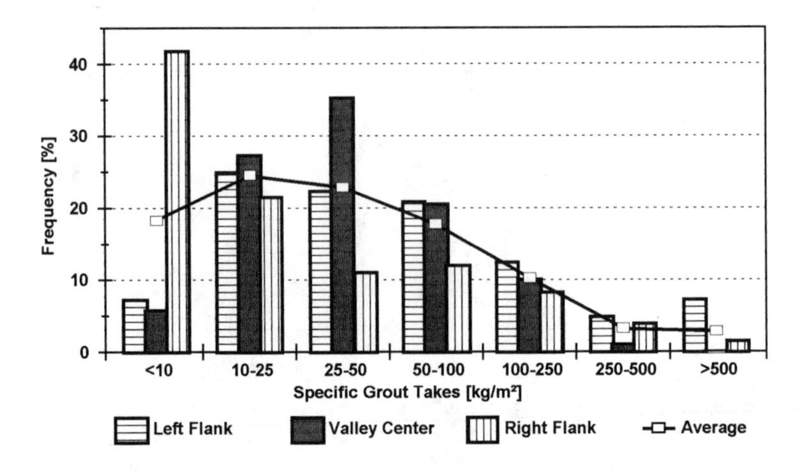

Figure 2: Frequency distribution of specific grout takes

4. Concepts of Hydrogeological Monitoring

Given the geological and hydrogeological conditions of the site, the grout curtain had to serve several purposes: a) to reduce seepage losses to an economically acceptable level, b) to keep uplift, seepage gradients and seepage velocities at magnitudes which are safe for the dam and appurtenant structures and especially below levels likely to cause erosion damage to the dam or the foundation rock. With the wide variation in quality of the foundation rock and its quite erratic grout response, the originally envisaged conventional closure criterion, specifying final Lugeon values of 1 to 2 l/min/m, was soon found insufficient for the assessment of adequate completion and performance of the curtain. Therefore, the trend in Lugeon values and specific grout consumption over the consecutive phases of grouting as well as the progressive changes of high takes percentiles were also taken into consideration. The latter criteria gave useful guidance for grouting technology but did not provide much information on the actual hydraulic performance of the curtain, as well as such related phenomena as downstream seepage gradients, underseepage, and the requirements and admissibility of drainage. To keep these important phenomena under control, a hydrogeological monitoring system was needed.

5. Methods and Procedures of Hydrogeological Monitoring

The most straightforward procedure would have consisted in the measurement of seepage flows. However, there were two major shortcomings: 1. To measure the total seepage flow, the direct discharges from the reservoir had to be interrupted. This required consent of the downstream countries Syria and Iraq and, therefore, only four such measurements could be taken. 2. The bulk seepage flow displays the overall performance but it does not provide information on the seepage paths and on the performance of specific segments of the very extensive curtain. These shortcomings were partially mastered by monitoring of the partial seepage flows from drainage galleries in the dam abutments and at the spillway and powerhouse. However, under the continuing construction works, these measurements had been limited in accuracy. Therefore, priority was given to piezometric monitoring. To eliminate ambiguities in the interpretation of the piezometric observations and to obtain quantified estimates of seepage flow, the piezometric monitoring was complemented by hydrochemical monitoring and by tracer tests.

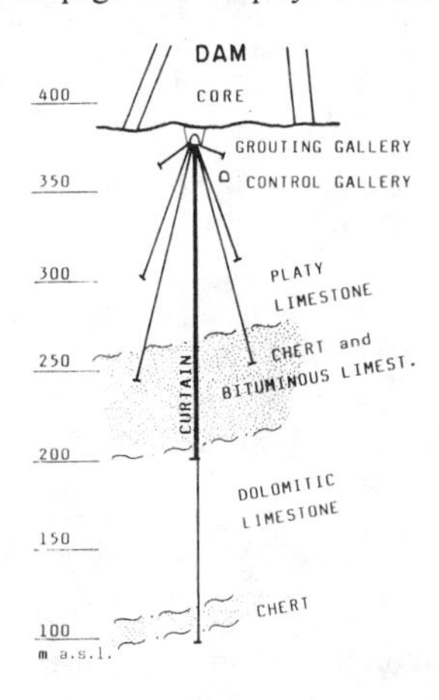

Figure 3: Typical arrangement of curtain piezometers in the valley floor

For the verification of quantitative aspects, simulation by digital groundwater models was carried out.

5.1 The Piezometric Monitoring System

Monitoring section were spaced at 150 m along the curtain. Standpipe piezometers, in upstream/downstream pairs, were placed on intervals in elevation of 50 to 70 m, and at a distance of 20 to 40 m from the centerline of the curtain (Figure 3). Additional piezometers were installed in the foundation of the core and beneath the concrete structures. There are currently about 300 piezometers in the dam foundation and along the grout curtain, 11 piezometers reach into the deep, semi-confined aquifer, and another 43 piezometers cover the abutments and the area downstream of the dam. Readings were taken on weekly intervals.

The basic routine for data reduction consisted of plotting hydrographs for all piezometers with the aid of a data base. The hydrographs provided useful indications of anomalous or unsatisfactory developments during the construction of the grout curtain and in this way helped substantially in quality control and the adjustment the work to the encountered conditions. When the interpretation became difficult, but also for semi-annual comprehensive evaluation, the piezometric contours were traced on sections parallel and transverse to the curtain. These contours were plotted in percent of total lake-tailwater head difference to allow direct comparison for different stages of lake filling. Figure 4 shows, how an incompletely sealed karst channel was identified

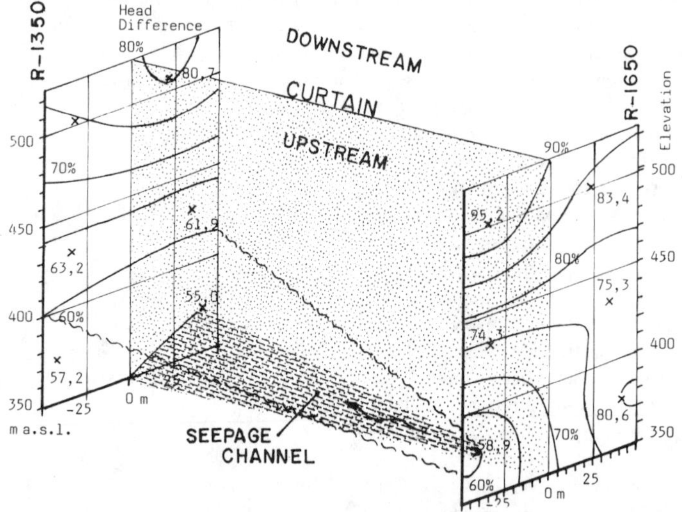

Figure 4: Low head at elevation 350 m indicates residual open karst channel crossing the curtain between chainage R-1650 upstream and R-1350 downstream. Head is stated in % of reservoir-tailwater difference

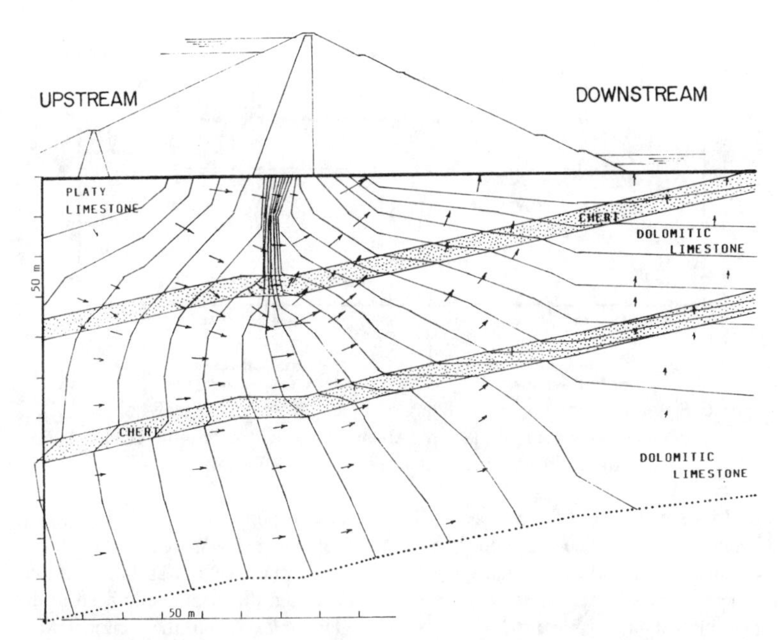

Figure 5: Model simulation of equipotential lines and flow pattern in the dam foundation

on the piezometric cross sections in the right abutment. For more quantified interpretation, mathematical models were run, calibrating the hydrogeological parameters against the observed head distribution (Figure 5). In this way the effective permeability of the curtain and the rates of cross seepage and underseepage could be estimated.

The piezometric contours on upstream and downstream envelopes of the curtain indicated the paths of seepage flow while the contour plot of head difference across the curtain (Figure 1) offered a comprehensive overview of the hydraulic performance of the curtain and helped to locate less effective spots.

Since piezometric readings were taken during reservoir filling, the piezometric response could be plotted against reservoir level. The slope of these plots was named "piezometric response coefficient". Determination of the coefficient gave additional insight into the hydrogeological performance of curtain and aquifers. The response coefficient is a direct function of the downstream portion of hydraulic resistance along the flow line passing at the piezometer. For instance, an abrupt drop in the response coefficient at downstream piezometers clearly marked a major hydrofracturing event in the confining layer of the deep aquifer (Figure 6). The response coefficients were also

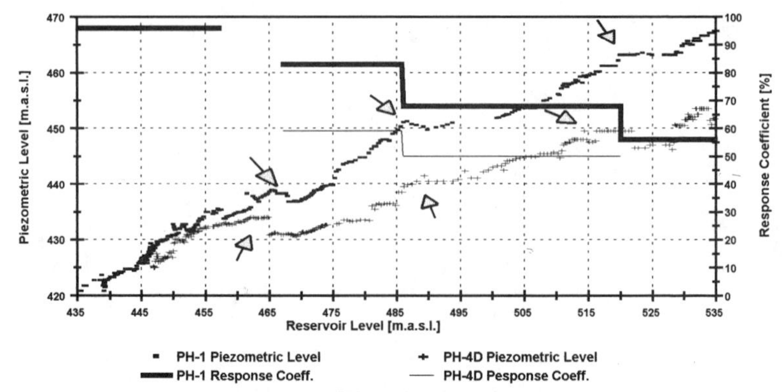

Figure 6: Response during reservoir filling of 2 piezometers in the semi-confined aquifer downstream of the dam. Arrows mark hydrofracturing events. The slope of the hydrograph is termed "response coefficient".

used to assess the need for drainage. When the first responses at the commencement of impounding had been obtained, projections of future downstream water levels were. made and a system of drainage galleries for both abutments was implemented. Furthermore, the water levels of the phreatic aquifer and the piezometric head in the semi-confined aquifer were traced in plan view (Figure 7). Again, the contours were checked by model simulation, to decide to what extent their configuration resulted from geometrical boundary conditions and to what extent it was indicative of preferential flow paths and local inefficiencies of the curtain.

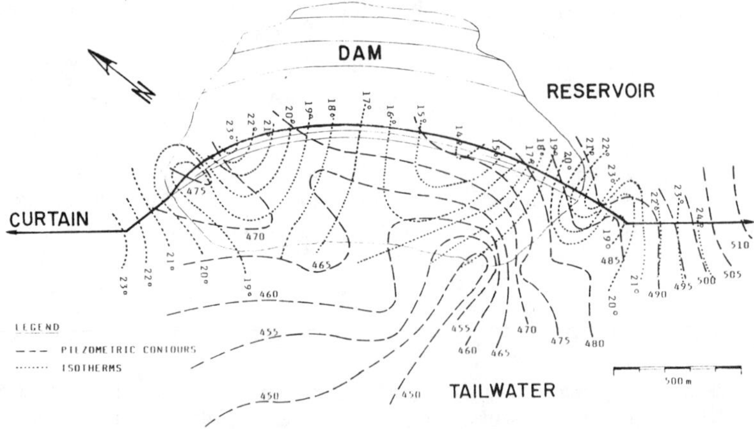

Figure 7: Piezometric contours [m.a.s.l.] and isotherms [°C] in semi-confined aquifer

5.2 Tracer Tests and Water Quality Monitoring

In the early stages of reservoir filling, the piezometric monitoring indicated several flaws in the curtain, especially on the right abutment. As a cross check on the piezometric interpretation, tracer tests with fluoresceine and rhodamine were carried out which determined for one of the suspect spots a pore velocity of the seepage flow in excess of 0.1 m/s. For comparison, the seepage flow pore velocity across the curtain in the valley floor is kept in the order of 10^{-5} m/s. With hydraulic gradients in the range of 5 to 10 and an estimated residual effective porosity of 5%, this low pore velocity would imply a permeability of the curtain corresponding to about 1 Lugeon unit. Two more tracer tests were made on later occasions. These tests proved useful but had the inconvenience of being time consuming and of causing a sustained contamination of the aquifers with the tracers. For these reasons, they could only be run on long intervals.

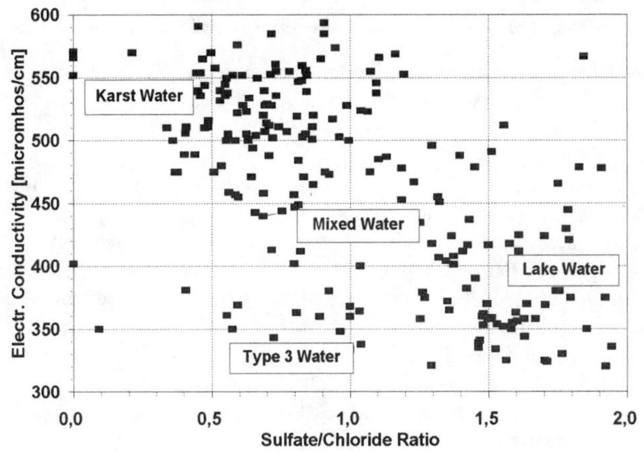

Figure 8: Hydrochemical characterization diagram for groundwater at the Ataturk site

Fortunately, hydrochemical and isotope monitoring had already been initiated at an early stage of the project. This program was later complemented by close monitoring of water temperature and electrical conductivity on short intervals. Chemical and isotope composition of the karst groundwater differed significantly from the surface water in the river and the reservoir (Figure 8). Thus, the water quality parameters could be used as natural tracers to monitor the movements of reservoir seepage water in the underground. Figure 9 clearly demonstrates the arrival of reservoir seepage water at a piezometer at the downstream toe of the dam not later than 3 months after commencement of the impounding. The water quality monitoring provided the first clear evidence that there was a direct hydraulic connection between the reservoir and the deep, semi-confined aquifer. On this basis, the flow velocities could be computed and the flow rates could be estimated.

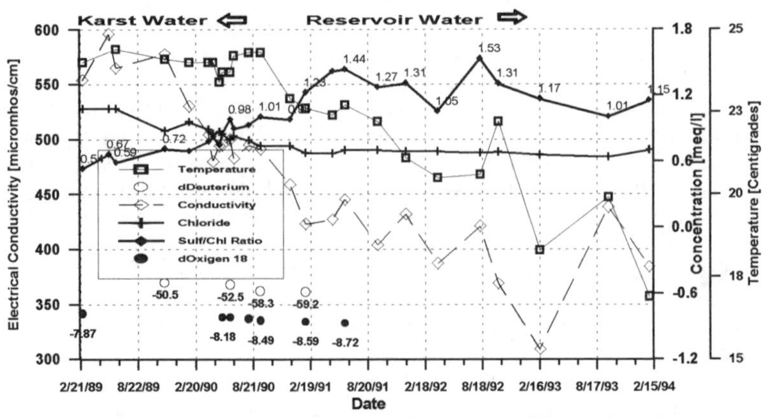

Figure 9: Change in composition of water during reservoir filling at deep piezometer at the downstream toe of the dam. Isotope delta values are stated in o/oo, referred to Standard Mean Ocean Water.

The capacity of the laboratory put a constraint on the rate and number of isotope and hydrochemical sampling. This constraint did not exist for the temperature and conductivity measurements which were taken in large number on monthly intervals. For evaluation, the data were plotted for each observation point against time. The interpretation of these diagrams benefited from the seasonal temperature oscillations in the lake. The propagation of the temperature extremes could be tracked in the aquifer and thus served as physical tracer pulses permitting estimates of seepage velocities between the piezometer pairs which straddle the curtain (Figure 10). In some cases the simple identification of particularly cold water at the curtain helped to locate preferential seepage paths.

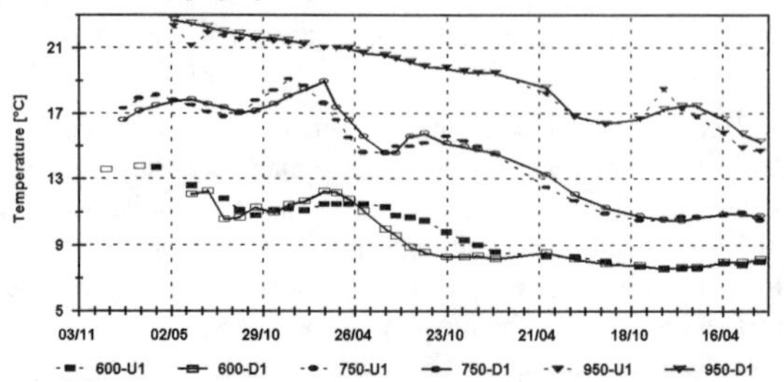

Figure 10: Temperature variation [°C] in time at curtain piezometers.

Plotting of groundwater isotherms in plan view and sections complements the piezometric evaluation and helps in the interpretation of the groundwater flow regime.

For instance (in Figure 7) a sink in the piezometric contours for the deep aquifer marks the area where the confining layer suffered hydrofracturing and where deep underseepage was being released. The isotherms, on the other hand, delineate the main paths along which the cold reservoir water flowed to the sink. As demonstrated in Figure 11 for the upstream envelope of the curtain, the isotherms were correlated with the piezometric conditions evaluated in Figure 1. The water temperature was depressed on both sides of the valley center where also the head difference indicated a comparatively low efficiency of the curtain. Additionally, the high temperatures in the vicinity of both abutments implied that the regional groundwater flow from the karst aquifer still persisted, despite a rise of 150 m in the discharge level. This circumstance reflected favorably on the regional water tightness of the reservoir. Directly adjacent to the upwelling warm karst water, a cold spot in the right flank at elevation 300 m marked an active karst channel, conveying reservoir seepage, and which to date still remains to be sealed (Figure 11).

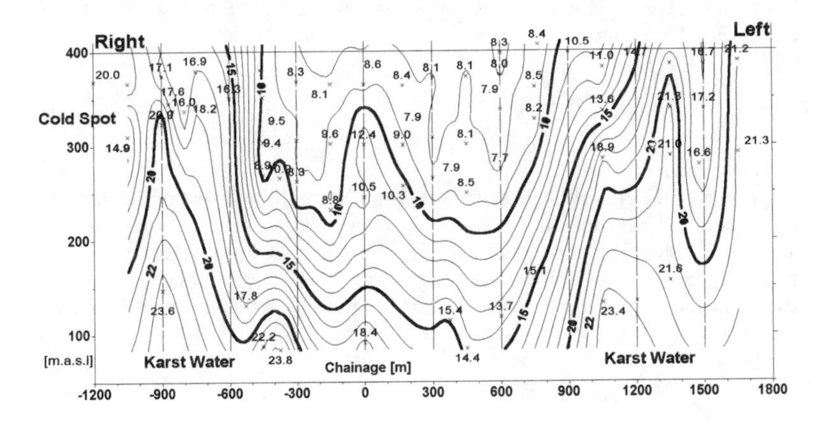

Figure 11: Groundwater Isotherms [°C] on upstream envelope of grout curtain.

6. Conclusions

Construction of a grout curtain in heterogeneously karstified and partially weak and erodible rock cannot be guided by simple, conventional closure criteria. Comprehensive and circumspect monitoring has to be made to assure that the curtain is adequate in extension and hydraulic performance and that the drainage system achieves its purpose without undue increase in seepage flows. Owing to the specific conditions prevailing at the Ataturk dam site in Turkey, the use of direct seepage monitoring was severely restricted. Therefore, reliance had to be placed mainly on piezometric monitoring, complemented by tracer methods, introducing artificial tracers but also utilizing physical and chemical properties of the seepage water as natural tracers. These techniques permitted the identification of spots in the curtain where improvement was required. They also helped to detect changes in the performance of the curtain and in the aquifers at the dam site in general. Such changes are considered typical for karstic rocks under severe hydraulic loading. They have to be followed carefully to assure that appropriate action is taken in due time before serious deterioration of the foundation takes place by erosion. Projections made at the commencement of reservoir filling had arrived at potential seepage losses of 20 m^3/s. With the subsequent extension and improvements of the curtain, the most recent measurement confirmed only about half of this earlier estimate. About 40% of the underseepage was passing through the deep, semi-confined aquifer. Thus, the performance of the grout curtain at the Ataturk dam site reached an acceptable standard. However, the instabilities in the karst aquifer, as evidenced, for instance, by recurring changes in piezometric response coefficients, impose the need for continued observation, and re-grouting or strengthening works will probably have to be contemplated in the future.

Acknowledgment
The owner of the project, State Hydraulic Works of the Government of the Republic of Turkey, had contracted the technical assistance of the Ataturk Engineers Joint Venture. Construction works had been carried out by ATA Insaat. The authors want to acknowledge the contribution to the construction and monitoring of the grout curtain by the engineers and geologists working for the Turkish government, for the Engineers Joint Venture and for the general contractor.

References

Ataturk Engineers Joint Venture. Miscellaneous unpublished grouting and monitoring reports.

Basmaci, E., 1991: Foundation behaviour of Ataturk dam on first filling. Proc. 17th Congr. ICOLD, Q. 66, R. 57, pp. 1051-1062

Deere, D. U., 1982: Cement-bentonite grouting for dams. Proc. Conf. Grouting in geotechnical engineering. W. H. Baker ed., ASCE, pp. 279-300

Compaction Grouting in a Canyon Fill

Russell C. Lamb[1], Member, and Daniel T. Hourihan[2]

A commercial building overlying a maximum 23 m (75 ft) deep canyon fill had experienced differential movement of 15 cm (6 in) over a 10-year period. Geotechnical investigation of the site disclosed that fill soil below a depth of about 9 m (30 ft) had an average relative compaction of 83 percent and that soil below a depth of 15 m (49 ft) was undergoing moisture-induced collapse. A compaction grouting program was implemented to stabilize the loose fill deposits and minimize future building movement. The program included injecting 832 m^3 (29,386 ft^3) of grout within 76 holes that were spaced on a 4.6 m (15 ft) square grid. Due to the existence of a functioning canyon subdrain within the treatment zone, a grout containment ceiling was constructed to avoid disrupting the subdrain operation. The program was monitored to assess compliance with specifications and included tests for the grout composition, consistency and injection rate, and observation of the subdrain operation. The completed program was evaluated by comparing grout volumes, grout to fill volume ratios, and pressures from production and verification grout holes, and by computing the increase in the average relative compaction of the fill after grouting. It is concluded that the grouting program increased the average relative compaction of the fill mass below 9 m from 83 to 88 percent, and that the containment ceiling effectively protected the subdrain.

---

[1] Senior Engineer, Stoney-Miller Consultants, Inc., 14 Hughes, Suite B101, Irvine, CA 92718

[2] Vice President, Denver Grouting Services, 7404 Trade Street, San Diego, CA 92121

Introduction

A commercial building located on a site within an industrial park in San Diego, California began to show notable distress approximately seven years after construction. The building is centrally located on the site and surrounded by slope, landscape, parking and drive areas (see Figure 1). The building consists of one-story tilt-up concrete panels and a concrete slab-on-grade floor. Wall and roof support are provided by continuous and isolated column footings, respectively. Building distress and movement was concentrated in the western one-third of the building, which was tilted down in a south and west direction. Distress to the building included cracks and separations in the floor slab, rotation and vertical cracking of wall panels, and distress to inset connections between panel walls. Level surveys performed on the floor slab and tops of the panels indicated that a maximum differential building movement of 15 cm (6 in) and maximum angular distortion of 1/120 had occurred since construction.

The construction documents indicated that the site was graded using techniques commonly employed in the southern California area. Pregrading topography at the site consisted of a southwest sloping hillside and a south draining canyon as shown in Figure 1. Total relief across the site was about 30 m (100 ft). Grading was completed in the early 1980's and 2 to 3 years prior to building completion. Grading operations consisted of removing vegetation and compressible materials from planned fill areas, installing a canyon subdrain, and placement and compaction of fill soil. The subdrain is a 20 cm (8 in) diameter, perforated metal pipe encased in 1 m^3 of 13 mm gravel per meter along the pipe. Fill was to be placed onto properly prepared competent materials, and compacted in order to achieve a minimum of 90 percent of the modified Proctor maximum dry density in accordance with ASTM D1557. The approximate maximum depth of fill placed was 23 m (75 ft) located near the southwest corner of the lot (see Figure 1).

A review of the geotechnical report of grading indicated that fill soil located about 9 m (30 ft) or more below pad grade was not adequately tested, and thus possibly not adequately compacted. A comparison of actual ground movement to that of well-compacted fills comprised of similar materials indicated that the fill was not performing normally with respect to either the magnitude or time rate of settlement. A subsurface geotechnical investigation of the site was conducted to determine the cause of building distress and movement.

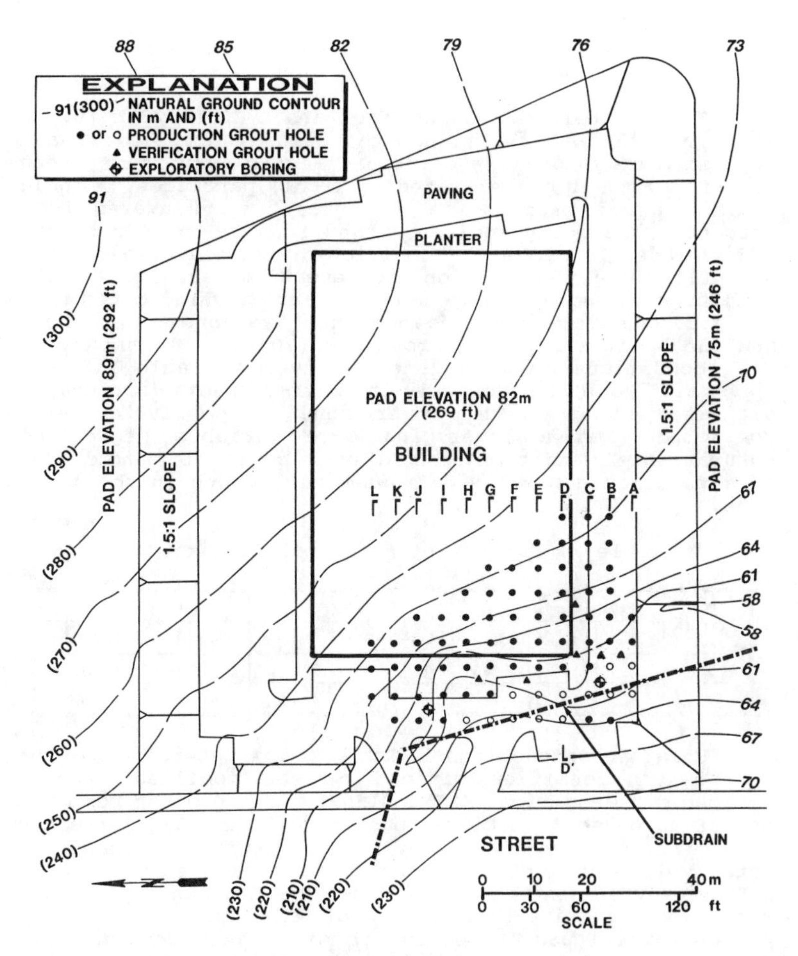

Figure 1. Site and Grouting Program Plan

Geotechnical Investigation

A geotechnical investigation of subsurface conditions was conducted, which included excavating and sampling of two bucket auger borings (see Figure 1), laboratory testing, and analyses of the results. Field work included undisturbed sampling or downhole sandcone testing (ASTM D1556) at 1.5 m (5 ft) vertical intervals in both borings. Relatively undisturbed samples of the fill soil were collected by pushing a pitcher tube sampler with the drill rig kelly bar. Both borings were excavated through the fill soil into bedrock.

Fill soil depths ranged from about 20 to 23 m (67 to 75 ft). Fill materials in both borings were similar and consisted of sandy lean clays and lean clayey sands, both with and without gravel-sized (> 19 mm) particles. Liquid limits ranged from 28 to 46 percent, and averaged 37 percent. Plasticity indices ranged from 11 to 26 percent, and averaged 19 percent. Specific gravities averaged 2.65. All fill was founded on competent materials. Fill conditions in both borings were similar, and indicated that the fill mass could be divided into two zones, one above and one below a depth of about 9 m (30 ft). A summary of fill conditions in each zone is given in Table 1. The upper fill zone is moderately- to well-compacted and nearly saturated, whereas the lower zone is poorly-to well-compacted with a lower and more variable degree of saturation. In addition, the relative compaction and degree of saturation generally decreased with depth in the lower zone.

Table 1. Summary of Fill Soil Conditions

Fill Depth (m)	Relative Compaction (%)			Saturation (%)			Percent Gravel		
	Low	High	Average	Low	High	Average	Low	High	Average
0-9	83	95	88	84	100	95	0	28	11.4
>9	71	92	83	62	100	82	0	15	6.9

Relative compactions in Table 1 were determined as the insitu dry densities divided by the modified Proctor maximum dry densities (ASTM: D1557). The Proctor test was performed on soil particles passing a 19 mm (3/4 in) sieve (i.e., it excludes gravel and larger particles). Thus, the total volume and the weight of the solids must be reduced by the gravel loss when determining the dry density and the resulting relative compaction of the soil. When the percent gravel was 10 percent or more, rock-corrected dry densities were determined using the method given in ASTM D4718 and were used to determine the relative compaction.

It was concluded that significant portions of the fill soil below a depth of about 9 m (30 ft) were not well-compacted. Collapse testing of fill soil samples showed that damaging future settlement of the building could occur due to moisture infiltration into the fill below a depth of about 15 m (50 ft). Remediation of the site and building was deemed necessary to achieve an acceptable level of future performance.

Stabilization Design

Compaction grouting was selected as the best alternative for site remediation. The objective of the grouting was to densify the fill soils supporting the building to near that required by the original grading specifications in order to minimize future settlement. A significant constraint was to avoid impacting the operation of the existing canyon subdrain. Due to the competency and high degree of saturation found within the upper fill zone (< 9 m or 30 ft), it was felt that grouting of this zone would result in little to no fill densification. However, grouting was to be performed within this zone to verify fill conditions. The lower fill zone (> 9 m) was found to have a sufficiently low degree of saturation so that grouting could adequately densify the loose fill soil, particularly for the collapsible fill soil located below a depth of about 15 m (50 ft).

The plan layout of the grouting program, shown in Figure 1, was selected based on observed building damage and distortion, knowledge of subsurface conditions, and the results of a compaction grouting test program conducted at an adjacent site situated over the same canyon fill. Figure 1 shows the production grout holes as closed and open circles; the closed circle holes were to be drilled into bedrock, whereas the open circle holes were to be drilled to about 2 m above the invert of the canyon subdrain. All holes were to be grouted from the bottom-up. The southern and western limits of grout holes were selected by extending a plane from the building limits to intercept with the bedrock surface as shown for Grout Line D in Figure 2. The northeastern limit of the program was selected to parallel the 70 m (230 foot) natural ground contour, i.e., the approximate top of the poorly-compacted fill soil zone.

The hole spacing, equipment, grout materials and consistencies, injection rates, and injection criteria for use during grouting were selected by the authors based on the results of the test grouting program. This program was devised and implemented using the compaction grouting technology presented in Brown and Warner (1973), Warner and Brown (1974), Warner (1982,1992), and Warner, et al. (1992).

Grouting Program

The grouting plan consists of 76 production grout holes spaced on a 4.6 m (15 ft) square grid pattern, and 6 verification grout holes located as shown on Figure 1. The production hole locations are shown by closed and open circles, and verification hole locations are shown as

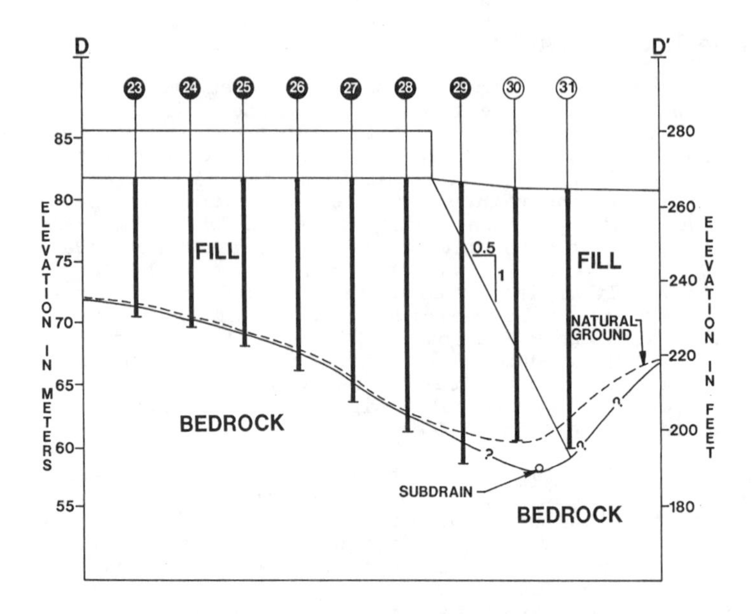

Figure 2. Cross Section Through Grout Line D

triangles. Grout holes were typically drilled vertically 0.9 m (3 ft) into competent materials as determined by difficult drilling and estimated depth to the pre-graded ground level. Open circle holes, however, were drilled to about 2 m above the subdrain invert. All holes were grouted from the bottom-up using 1.5 m (5 ft) vertical stages, except for the open circle holes; in these a 1.5 m thick grout containment ceiling was first constructed at the base by injecting a maximum of 0.45 m^3 (16 ft^3) of grout per 0.3 m (1 ft) stage. The ceiling grout was then allowed to set for at least 24 hours prior to additional grouting.

All grout was placed using an injection rate of 0.94 Liter/sec (2 ft^3/min). Grouting at any one stage usually continued until the first of the following:

1. A 1.3 mm (0.05 in) vertical uplift occurred at the surface adjacent to the grout point.
2. System pressure at the probe header was:
   a. 2.41 MPa (350 psi) when the probe was less than 10.7 m (35 ft) below grade.
   b. 4.14 MPa (600 psi) or greater when the probe was 10.7 m or more below grade.
3. A maximum of 4.25 m^3 (150 ft^3) of grout was placed within any 1.5 m (5 ft) vertical stage

for holes located on the south and west boundaries of the grid.

A summary of the grouting program is given in Table 2, which lists the Grout Hole No., total grouted depth, total grout volume, and sequence of grouting. Grout hole numbers given in Table 2 are keyed to Figure 1. The letter designation identifies the Production Grout Line A through L shown in Figure 1. Production grout holes are numbered sequentially starting at the east of Grout Line A and proceeding west, then to the east of Grout Line B and proceeding west, etc. Verification grout holes, V-1 through V-6, are numbered sequentially moving in a similar east to west, south to north direction.

Table 2. Grouting Program Summary

Grout Hole No.	Total Depth (m)	Total Grout Volume (m^3)	Sequence
A-1	20	9	15
A-2	24	15	12
A-3	23	15	16
A-4	23	8	6
B-5	12	5	26
B-6	15	5	31
B-7	14	4	27
B-8	16	5	32
B-9	18	14	28
B-10	20	12	34
B-11	21	34	23
B-12	21	14	18
B-13	20	19	7
C-14	13	3	33
C-15	14	12	37
C-16	15	5	35
C-17	16	8	38
C-18	18	15	36
C-19	22	26	43
C-20	21	18	39
C-21	21	15	11
C-22	20	31	1
D-23	11	4	53
D-24	12	8	49
D-25	14	3	67
D-26	16	10	52
D-27	16	7	56
D-28	21	10	50
D-29	23	4	46
D-30	21	8	14
D-31	21	13	3
E-32	12	1	75
E-33	13	4	68
E-34	15	3	74
E-35	18	14	51
E-36	20	7	66
E-37	23	18	44
E-38	21	22	17
E-39	21	15	2
F-40	13	3	76
F-41	15	6	61
F-42	18	2	65

Grout Hole No.	Total Depth (m)	Total Grout Volume (m^3)	Sequence
F-43	20	13	55
F-44	22	8	47
F-45	20	10	20
F-46	20	22	4
G-47	12	6	59
G-48	14	2	73
G-49	17	8	60
G-50	20	7	64
G-51	22	28	45
G-52	22	13	25
G-53	20	19	5
H-54	14	3	62
H-55	15	3	72
H-56	19	11	54
H-57	21	4	48
H-58	22	26	29
H-59	18	18	9
I-60	14	3	63
I-61	16	5	69
I-62	20	16	41
I-63	22	11	19
I-64	21	13	13
J-65	13	2	70
J-66	15	7	57
J-67	15	14	30
J-68	18	16	22
J-69	19	22	10
K-70	14	3	71
K-71	14	13	40
K-72	16	10	24
K-73	18	19	8
L-74	12	3	58
L-75	12	6	42
L-76	15	14	21
V-1	25	2	77
V-2	23	3	81
V-3	22	1	78
V-4	17	2	82
V-5	23	4	79
V-6	22	2	80

The total quantity of hole drilled was 1341 and 132 lineal meters (4400 and 433 ft) for production and verification grouting, respectively. Total grout volumes were 832 and 14 m^3 (29,386 and 495 ft^3) for production and verification grouting. Injection pressures typically ranged from 1.5 to 4.14 MPa (217 to 600 psi).

Equipment

Grout holes were drilled using air percussion or water-rotary methods. Grout materials were mixed in a volumetric batcher. Grout was injected using positive-displacement pumps as discussed by Warner (1992). A Putzmeister TS-2015 pump was generally used for injection, but a Thomsen 2002-D pump was employed intermittently throughout the project to allow for equipment maintenance.

Materials

The grout consisted of silty sand, cement, and water. The gradation of the silty sand was near the finer limit of that recommended by Warner and Brown (1974). The average weight composition of the grout was 1:0.27:0.1 (silty sand:water:cement). The silty sand was supplied by FST Sand and Gravel from their site in San Juan Capistrano, California. The cement was Type I. Potable water was supplied from an on-site fire hydrant.

Monitoring

Grouting operations were observed and tested full-time by a representative of the first author. Field testing of the grout included companion slump (ASTM: C143) and moisture content tests (ASTM: D2216). Grout slumps were generally maintained between 25 and 50 mm, for which water contents ranged from 23 to 26 percent. An injection rate of 0.94 Liter/sec (2 ft^3/min) was used and was periodically determined by filling a container of known volume and recording the time for filling. Both drilling and grouting logs were maintained as recommended by Warner and Brown (1974).

The operation of the canyon subdrain was observed from the base and top of a manhole located on the downhill property and about 15 m (50 ft) south of Grout Line A (see Figure 1). Periodic observation from the surface during grouting showed that water was always flowing in the subdrain. Moreover, a 2.5 cm depth of water was observed flowing in the subdrain pipe both prior to and after grouting.

Site, building, and floor slab elevations within the grouted area were determined prior to grouting, and at about one month and ten months after grouting. Elevation differences between pre-grouting and one month post-grouting ranged from +18 to -9 mm (+0.71 to -0.35 in), and averaged +4 mm (+0.16 in). Elevation differences between one month and ten months post-grouting ranged from ±6 mm (±0.24 in), and averaged +1.3 mm (+0.051 in). Most of these differences were within survey accuracy (±3 mm).

Evaluation Of Results

The results of the grouting program were evaluated by three methods to aid in assessing the degree of ground improvement. The first method consisted of comparing grout volumes and the ratio of grout to fill volumes during production (P) grouting to those during verification (V) grouting. The second method compares the variation of sustained pressures and grout volumes with depth for P and V grouting. Lastly, the increase in the average relative compaction of the fill soil after grouting is estimated.

Prior to comparing the grouting results, volumetric quantities of the fill and grout mass will be defined. The total fill volume (V_t) assumed affected by each grout hole is defined as a rectangular block with sides equal to the grout hole spacing and depth equal to the grout hole length within the upper (< 9 m or 30 ft) or lower (> 9 m) fill zone. The compactable fill volume (V_{tc}) is defined as the product of V_t and (1-PG), where PG is the percent gravel given in Table 1. Thus,

V_t (upper) = 14,296 m^3 = 5.05 x 10^5 ft^3;
V_{tc} (upper) = 12,666 m^3 = 4.47 x 10^5 ft^3;
V_t (lower) = 13,732 m^3 = 4.85 x 10^5 ft^3; and
V_{tc} (lower) = 12,784 m^3 = 4.52 x 10^5 ft^3.

The total grout volumes placed within the upper and lower fill zone during P grouting were 87 and 745 m^3 (3,073 and 26,313 ft^3), respectively; those during V grouting were 2.3 and 11.7 m^3 (81 and 413 ft^3), respectively.

Each of the six V holes drilled and grouted subsequent to P grouting are centered within a group of four P grout holes (see Figure 1). A comparison of results of grouting for the production and verification holes is given in Table 3. This table gives the average grout volume per vertical meter of hole and the grout volume (V_g) as a percent of the total compactable fill volume (V_{tc}) affected by grout injection. The upper zone has a relatively small average grout volume and grout to compactable fill volume percentage, as anticipated. Both quantities were reduced by at least two-thirds from P to V grouting. P grouting volumes within the lower zone were significantly higher than those in the upper zone. The average grout volume in the lower zone was reduced seven-fold from P to V grouting (see Table 3). A similar reduction in the grout to compactable fill volume occurred in the lower zone from P to V grouting.

Table 3. Comparison of P and V Grouting

Grout Hole Description	Depth (m)	Average Grout Volume (m^3/m)	V_g/V_{tc} (%)
Production	0-9	0.13	0.7
Verification	0-9	0.04	0.2
Production	>9	1.13	5.8
Verification	>9	0.15	0.8

The above comparisons tacitly assume that a reduction in grout volume from P to V grouting is evidence that the fill soil was densified, and that grouting cessation was the result of adequate sustained high pressures. A similar grout volume reduction could also occur due to surface lift.

To explore the cause of the reduced grout volumes, plots of sustained pressure and grout volume versus depth were prepared for three representative verification holes and their surrounding production holes. These are shown in Figures 3, 4, and 5 for the V2 group (i.e., Grout Holes V-2, 10, 11, 19 and 20), the V5 group, and the V6 group, respectively. In these, the sustained pressure scale is linear, whereas the grout volume scale is logarithmic. Open and closed symbols are used for each injection stage to

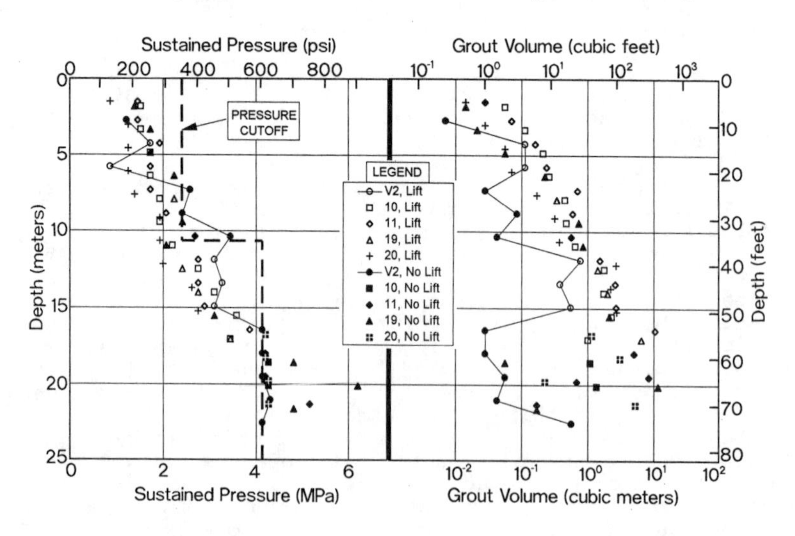

Figure 3. Pressures and Grout Volumes for Group V2

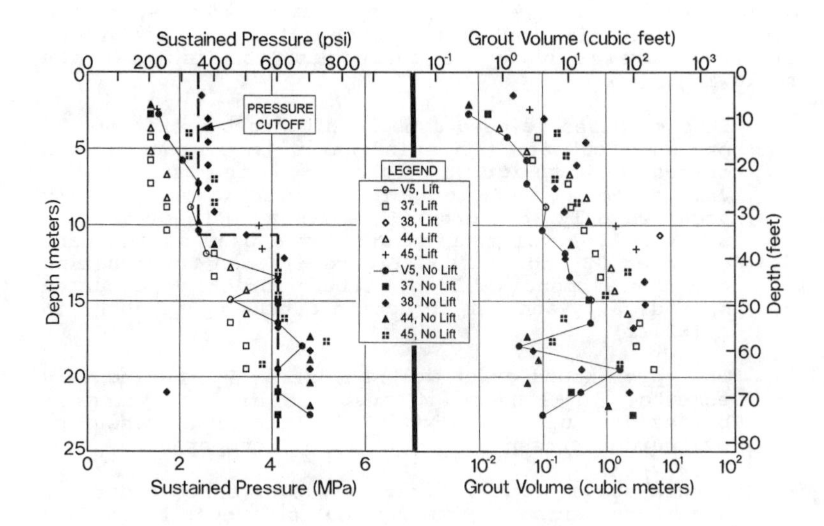

Figure 4. Pressures and Grout Volumes for Group V5

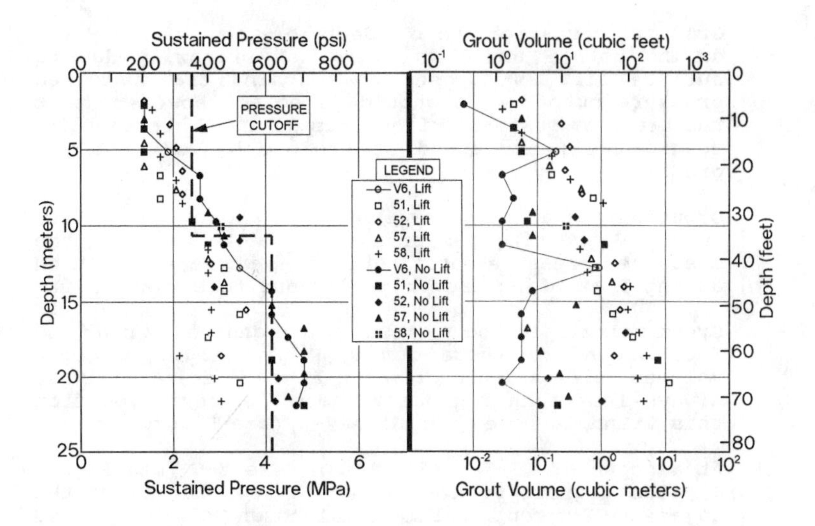

Figure 5. Pressures and Grout Volumes for Group V6

denote whether grouting cessation was due to surface lift or not due to surface lift, respectively. The sustained pressure criteria for grouting cessation is shown for clarity. Careful study of these results discloses the following:

- In some cases, closed symbols plot to the left of the pressure cutoff. This indicates that grouting was ceased due to grout leakage to the surface, a volumetric cutoff, or ground cracking. Leaking occurred only when grouting above a depth of about 3 m (10 ft). A volumetric cutoff was used at the base of Holes 38 and 45 (see Figure 4) when constructing the containment ceiling. Ground cracking occurred sporadically and for injection depths up to about 16 m (52 ft).

- The upper-bound grout volumes during P grouting vary somewhat log-linear with depth. This variation is believed to be a function of the sustained pressures, overburden pressure, and fill soil competency.

- Single-stage grout volumes vary over three orders of magnitude during P grouting, but only over two orders of magnitude during V grouting. The maximum single-stage grout volume ranges from about 7 to 12 m^3 (247 to 423 ft^3) during P grouting and 0.6 to 1.6 m^3 (21 to 56 ft^3) during V grouting.

- Grouting cessation above a depth of about 15 m (50 ft) during both P and V grouting was primarily due to surface lift as opposed to reaching the sustained pressure cutoff. It should be noted, however, that the trend of sustained pressures during V grouting is toward the upper bound of those achieved during P grouting.

- Grouting cessation below a depth of about 15 m (50 ft) during V grouting was primarily due to reaching the sustained pressure cutoff criteria, whereas cessation during P grouting was primarily due to surface lift.

- Grout volumes during V grouting are generally near the lower bound of those during P grouting. V grout volumes below a depth of about 15 m (50 ft) are almost always less than P grout volumes. A deviation from this trend is noted for Hole V-5 (see Figure 4).

- At a depth of 19.5 m (64 ft) in Hole V-5 (see Figure 4), the V grout volume was not a lower bound to the adjacent P grout volumes, although the specified sustained pressure was achieved. The likely cause of this is the volumetric cutoff (and resulting low sustained pressure) used to construct the containment ceiling for Hole 38. Fortunately this did not occur below a depth of about 15 m (50 ft) in any other V hole.

- At a depth of 12.8 m (42 ft) in Hole V-6 (see Figure 5), the V grout volume was near the lower bound of the P grout volumes, and the sustained pressure during V grouting was an upper bound for the adjacent production grout stages. A similar feature may be observed for the grout volume and pressure in Hole V-2 (see Figure 3). Since the fill soil above a depth of about 15 m (50 ft) was found to have a very low potential for future settlement at insitu density conditions, the simple reduction of grout volume from P to V grouting was considered acceptable.

- Sustained pressures during both P and V grouting vary somewhat linearly with depth. At this site, a sustained pressure of about 0.25 MPa/m of depth (10 psi/ft) was generally achieved during verification grouting. Based on experience, such pressures are generally achievable in well-compacted fills before significant surface uplift occurs.

The degree of ground improvement due to grouting may be quantified by estimating the increase in the average relative compaction of the fill. It can be shown that the ratio of the final average relative compaction (RC_f) to the initial average relative compaction (RC_i) is a function of the total compactable fill volume and the grout volume, i.e.,

$$\frac{RC_f}{RC_i} = \frac{V_{tc}}{V_{tc} - V_g} \tag{1}$$

Use of Eq. (1) requires that the total compactable fill volume (V_{tc}) remain constant. This is approximately true for this site since the average surface uplift at one month post-grouting was 4 mm (0.16 in). This represents an increase in V_{tc} of 6 m^3 (212 ft^3) which is negligible when compared to the original V_{tc} of 25,450 m^3 (8.99×10^5 ft^3).

For production grouting above a depth of 9 m, V_g = 87 m^3 and V_{tc} = 12666 m^3. Using Eq. (1) gives RC_f/RC_i = 1.007. Then the upper zone, which had an RC_i of 88 percent (see Table 1), has a RC_f of 88.6 percent. For production grouting below a depth of 9 m, V_g = 745 m^3 and V_{tc} = 12784 m^3. From Eq. (1), RC_f/RC_i = 1.062. The RC_i of the lower zone was 83 percent (see Table 1), thus the RC_f is 88.1 percent.

Conclusion

Based on the results of the completed grouting program, it is concluded that compaction grouting at the site effectively increased the relative compaction of the loose fill soil within the zone of influence of the building, and that the grout containment ceiling avoided disruption of water flow in the canyon subdrain. On the average, ground improvement was sufficient to result in a structural fill comparable to that required by current southern California grading codes with acceptable consolidation performance. It is also concluded that grouting of verification holes and analyses of the results is a valid means to assess whether adequate ground improvement has been achieved. The success of the program suggests that compaction grouting, employed similarly to that indicated herein, can be an effective and safe settlement remediation method in canyon fills with in-place subdrains.

Acknowledgements

The authors would like to thank Dr. Roy Borden, Mr. Hannes Richter, and Mr. Kevin Trigg for their review of and comments on the paper. Additional thanks are due to Mr. Jon Petitjean and Mrs. Katherine Dickinson for their preparation of final graphics, and Mrs. Jean Bennett for processing the paper.

Appendix I. References

Brown, D.R. and Warner, J. (1973). "Compaction Grouting", Journal Soil Mechanics and Foundation Division, ASCE, Vol. 99, No. SM8, pp. 589-601.

Warner, J. and Brown, D. R. (1974). "Planning and Performing Compaction Grouting", Journal Geotechnical Engineering Division, ASCE, Vol. 100, GT6, June 1974, pp. 653-666.

Warner, J. (1982). "Compaction Grouting - The First Thirty Years", ASCE Proc. Grouting in Geotechnical Engineering Conference, New Orleans, Louisiana, pp. 694-707.

Warner, J. (1992) "Compaction Grouting; Rheology vs. Effectiveness", ASCE Proc. Grouting, Soil Improvement, and Geosynthetics, New Orleans, Louisiana, pp. 229-239.

Warner, J., Schmidt, N., Reed, J., Shepardson, D., Lamb, R. and Wong, S. (1992). "Recent Advances in Compaction Grouting Technology", ASCE Proc. Grouting, Soil Improvement, and Geosynthetics, New Orleans, Louisiana, pp. 252-264.

Appendix II. Notation

The following symbols are used in this paper:

PG = percent gravel in soil expressed as a decimal;
RC_i = initial relative compaction of soil;
RC_f = final relative compaction of soil;
V_g = grout volume;
V_t = total volume of fill affected by grout injection; and
V_{tc} = total volume of compactable fill affected by grout injection.

Enhancement of Caisson Capacity by Micro-Fine Cement Grouting --A Recent Case History--

D.A. Bruce[1], P.J. Nufer[2], and R.E. Triplett[3]

Abstract

Pressure grouting techniques have been used for many years to enhance the performance of axially loaded caissons, either as a remedial technique, or as an integral step in the foreseen construction program. This paper describes the work undertaken with microfine cement to upgrade the capacity of existing caissons at an industrial facility in Jonesboro, AR. Details are provided of the test program, and specifically the performance of the test caisson, load tested before and after permeation grouting of the bearing stratum. Grouting produced an improvement in load-bearing capacity by over three times. The procedures used in the subsequent production program are also described.

The Problem

The Owner wished to expand a cereal production facility at his plant in Jonesboro, AR. This expansion involved adding a new structure outward from the exterior wall of the east side of the existing building. However, additional loading from this new structure was determined likely to overload four of the existing 14 foot long belled caissons under the common exterior wall, installed 6 years earlier.

----

[1]Vice President, Nicholson Construction Company, P.O. Box 98, Bridgeville, PA 15017 (412) 221-4500.
[2]Project Manager, Nicholson Construction Company, 1352 Union Hill Road, Alpharetta, GA 30201 (404) 442-1801.
[3]Construction Manager, Nicholson Construction Company, 1352 Union Hill Road, Alpharetta, GA 30201 (404) 442-1801.

Site Conditions and Caisson Load Test Performance Before Grouting

Data from boring E13 (Figure 1) indicated loose, moist, brown, clayey sand for the upper 12 feet, and N values of 3-10. Therebelow were a series of medium dense to dense brown and grey silty fine sands, with N values of 17 to 38. The water table was about 23 feet below the surface. Mechanical grain size analyses indicated a portion 6 to 13% finer than the #200 sieve.

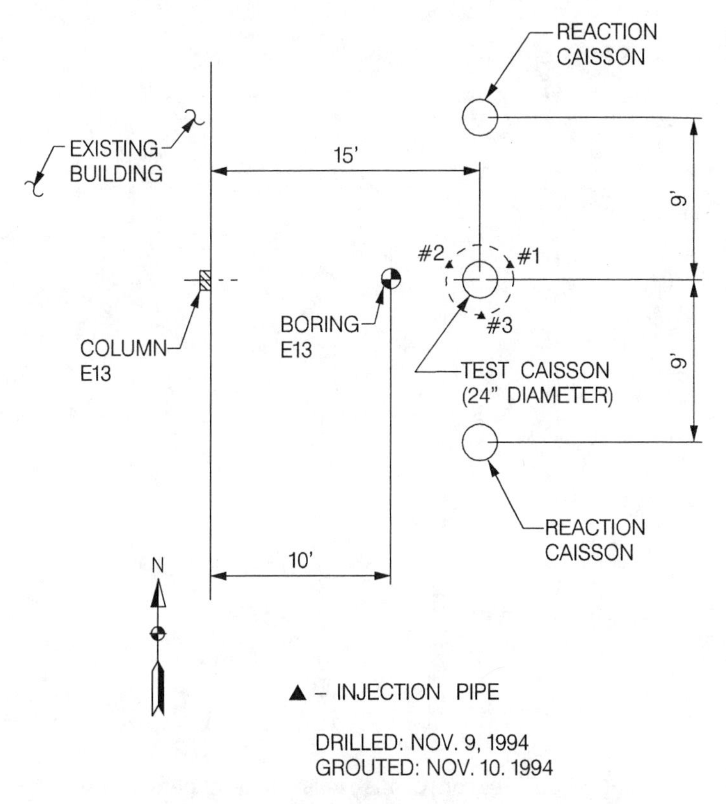

Figure 1. Plan of test caisson arrangement.

It was agreed to install and test a new caisson to better define the bearing capacity of the lower sand, the founding stratum for the existing piles. The test comprised a 24 inch diameter, straight shaft, reinforced concrete caisson, 14 feet long - the same depth as the existing belled caissons. It was cast inside a 24 inch diameter lubricated sonotube form to eliminate any skin friction component. Two reaction caissons were installed adjacent (Figure 2) to anchor the load frame.

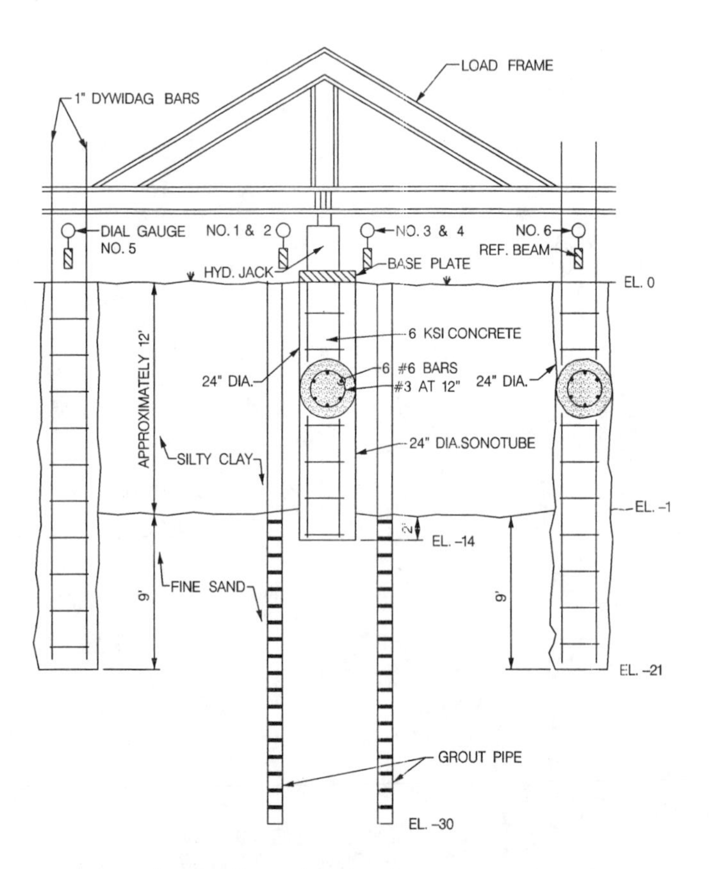

Figure 2. Section of test caisson arrangement.

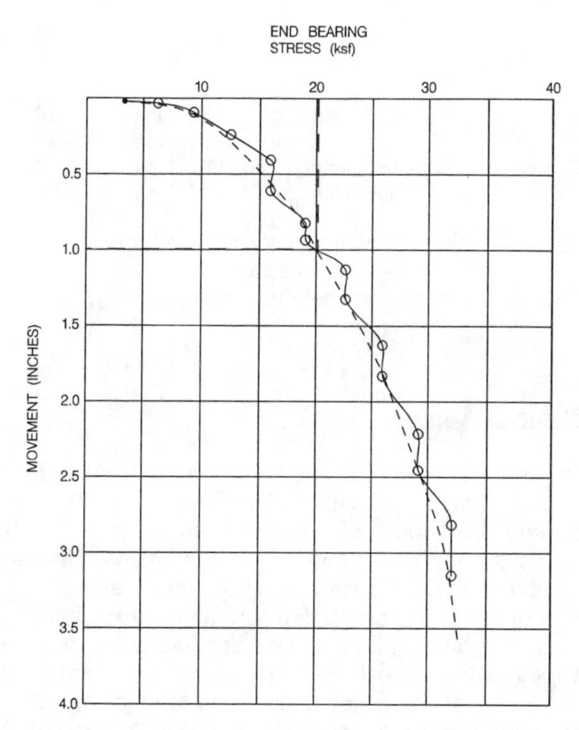

Figure 3. Movement/end bearing stress data for the test caisson, before grouting.

A compressive load test was conducted four days after caisson installation and in accordance with ASTM D-1143. This progressive incremental loading sequence showed that a test load of barely 40 tons could not be maintained with a creep rate less than 0.01 inch per hour. (Figure 3). The Engineer had indicated that the structure could withstand a further total movement of 1 inch. The analysis of the load test data confirmed the ultimate bearing capacity, at this displacement, was 20 ksf. Using a safety factor of 2, and so an allowable end bearing stress of 10 ksf, the Engineer calculated that the bearing capacity of the sand would be exceeded under four of the existing caissons, by factors of 11 to 20%, as shown in Table 1.

Caisson	Bell Diameter	Existing Allowable Load*	Required Future Load	Problem
E9	11 feet	950 kips	1133 kips	19% overload
E11	12 feet	1131 kips	1251 kips	11% overload
E13	12 feet	1131 kips	1357 kips	20% overload
E14	11 feet	950 kips	1102 kips	16% overload

Table 1. Details of existing caissons exceeding original allowable capacities
(*based on allowable bearing capacity of 10 ksf)

Concept of the Solution

Various options were considered, with regard to their impact on cost, time, practicality, likely effect on the existing structures, and likelihood of technical success. The most appropriate solution was agreed to be increasing the bearing capacity of the soil by some form of ground treatment. The contractor proposed a permeation grouting procedure, since this had met with success in similar conditions on other sites (Bruce, 1986), while consideration of the soil granulometry indicated that a microfine cement grout would be suitable. Given that a test caisson was already available, and that, by having tested it in advance of any grouting, it had provided a "baseline" performance level, it was elected to grout under this caisson, and to retest it as a way of demonstrating the effectiveness of the proposal.

Grouting of the Test Caisson

Three vertical grout holes, approximately 4 inches in diameter, were spaced equally around the caisson, and drilled to a depth of 30 feet. Much of the production work was to be conducted inside the building where low headroom restraints and tight controls on effluent and spoil had to be accommodated Therefore, the opportunity was taken to drill these holes, even though they were outside, with the short-mast crawler rig equipped with the 3 foot sections of continuous hollow stem flight augers, foreseen for the later production work.

Sheath Grout

5% Bentonite	9.4#
22.5 Gal. Water	188#
1 bag cement	94#
Specific Gravity	1.32
Marsh Funnel	80+ sec.
Bleed	<2%
Yield per bag of cement	3.5CF

Microfine Grout Mix #1 (W/C Ratio 2:1)

13.5 Gal. Water	110#
25 kg Microcem 900	55#
0.75% Eucon 37	0.4# (approximately 5 fluid ounces)
Specific Gravity	1.26
Marsh Funnel	28 sec.
Bleed	<1%
Yield per bag of cement	2.28 CF

Microfine Grout Mix #2 (W/C Ratio 3:1)

20 Gal. Water	165#
25kg Microcem 900	55#
0.75% Eucon 37	0.4# (approximately 5 fluid ounces)
Specific Gravity	1.19
Marsh Funnel	28 sec.
Bleed	-- (no data)
Yield per bag of cement	2.95 CF

Table 2. Details of grout mixes used.

Each hole contained a 1-1/2 inch diameter sleeved plastic pipe, with sleeves at 1 foot intervals over the lower 20 feet. The annulus grout was tremied in place, and consisted of a low strength, brittle, stable, water-cement-bentonite mix, as shown in Table 2.

The following day, injection through the sleeves was conducted with an inflatable double packer. Microcem 900 was selected as the microfine cement on grounds of technical and cost superiority. It was mixed in a colloidal mixer and pumped via a piston pump (Photograph 1). A water/cement ratio of 3:1 (by weight) was initially used, to verify rheological characteristics and ease of penetration. Manual records were maintained of the injection characteristics of each sleeve, and they quickly suggested a reduction in the water/cement ratio to 2:1.

Based on calculated grout travel distances and soil porosity, an upper limit volume of 1 bag per foot of grout hole was established, in the region between 29 and 15 feet below surface. Due to grout breaking out at the surface from injections into shallow sleeves, the total actually injected was 39 bags. Results from the fluid tests conducted on the various mixes are detailed in Table 2. Injection pressures ranged up to 225 psi and injection rates varied up to 1.7 cf/min. No upwards movement of the pile was recorded during grouting.

Photograph 1. Grouting underway around the test caisson.

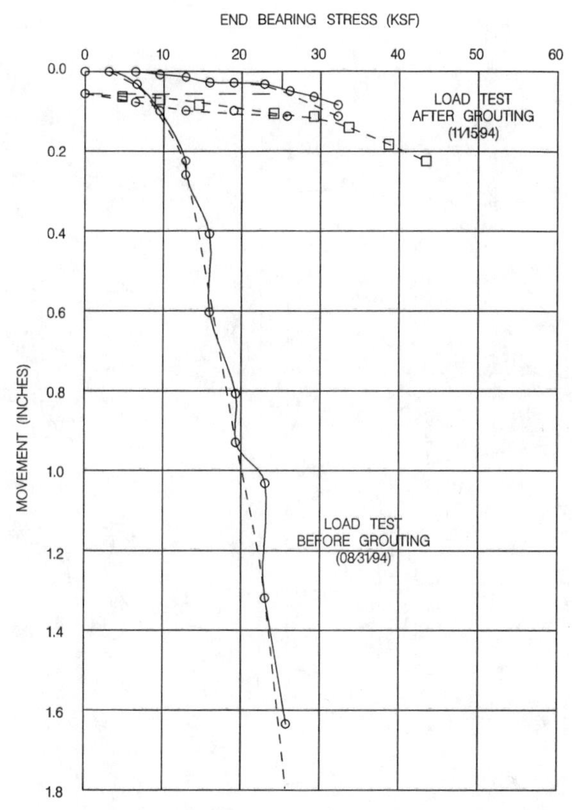

Figure 4. Movement/end bearing stress data for the test caisson, before and after grouting.

Caisson Load Test Performance After Grouting

Five days after grouting, the caisson was retested (Figure 4). The initial test load of 50 tons was held for 3 hours with a total pile head movement of 0.101 inches including 0.020 inches of creep. In the following 23 hours, a further 0.009 inches of creep was recorded. Upon unloading to zero, the permanent movement was 0.050 inches. Thereafter, the owner gave permission to reapply load up to the effective maximum capacity of the loading system, namely 70 tons. This was achieved with a total movement (from start of test) of 0.311 inches, and minimal additional permanent movement.

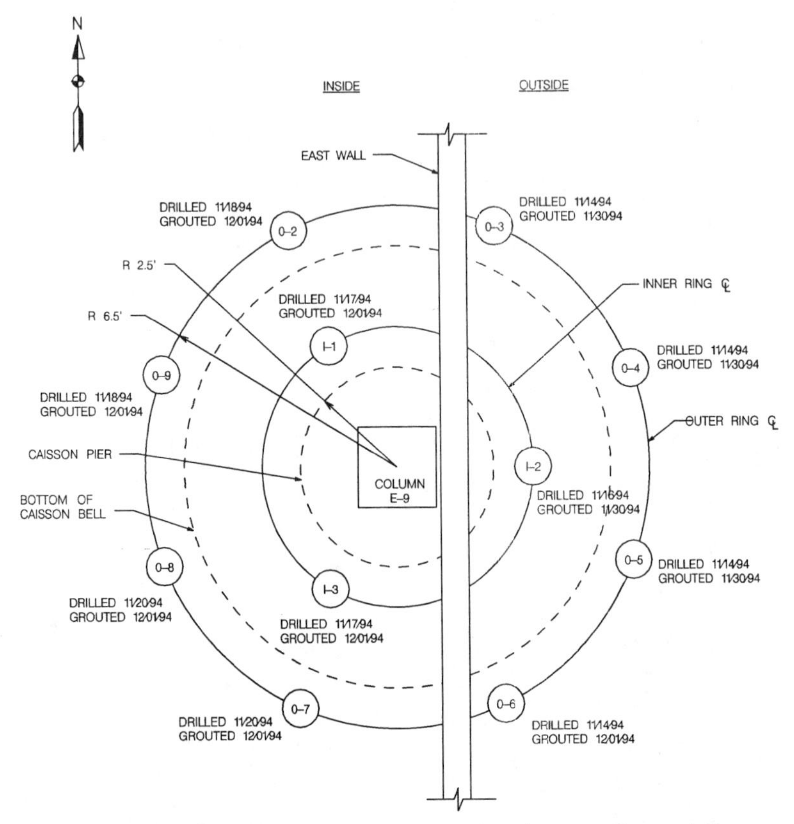

Figure 5. Typical grout hole layout plan for grouting existing caissons.

As a consequence, the allowable bearing pressure for piles in grouted sands was increased to 12.5 ksf at an estimated movement of 0.100 inch. (The tested capacity, at this movement was recorded as about 32 ksf).

Subsequent Treatment of Existing Caissons

Given the extremely positive results from the test caisson, it was decided to enhance the performance of these four existing caissons by similar methods. Eleven grout holes were drilled through and around each bell in concentric rings (Figure 5), to a depth of 15 feet below its base. Six of the eleven holes had to be installed and grouted from within the building (Photograph 2), necessitating precoring of the 6-12 inch thick floor slab. Drill methods and tooling had to be adjusted to accommodate the concrete and soil, but the grouting proceeded as for the test caisson with the 2:1 mix. Columns and floor slab were monitored during injection for uplift. Upon completion of grouting, the holes in the floor slab were backfilled with non-shrink grout.

Strong quality assurance/quality control measures were observed during each of the drilling and grouting processes, and detailed construction records were maintained. No accidents were recorded, the specialty grouting work was completed one week ahead of schedule in about 3 working weeks, and the overall facility expansion projection was not delayed as a consequence of having to conduct the caisson grouting program. Additional work of similar nature is to be carried out as the expansion program unfolds.

Photograph 2. Grout hole drilling in low headroom conditions for existing caisson 13E.

Final Remarks

Grouting to enhance the performance of caissons has been conducted for many years in Europe or by European contractors working abroad, and especially in the Middle East. More recently these techniques have begun to be applied in North America. This case history is a good example of the real benefits that a well conceived, properly managed and correctly executed grouting program can provide. Many technical points are described, and the effectiveness of the work clearly demonstrated. However, it is important to note that this project was allowed to proceed in a team atmosphere which fostered the design-build approach, and encouraged trust and respect amongst all the parties, to mutual benefit.

Acknowledgements

The authors wish to express their thanks to the various personnel from the Owner (Kraft General Foods), the Construction Managers and Engineer (Rust Engineering Company) and the Geotechnical Engineer (Anderson Engineering Consultants, Inc.) for their cooperation at every phase of this project. Microcem 900 is manufactured by Blue Circle Ltd. (United Kingdom), and distributed in the United States by Multiurethanes, Inc. (Buffalo, New York).

References

1. ASTM-D1143-81 (1981). Method of Testing Piles under Static Axial Compressive Load. Section 04, Vol. 04.08.

2. Bruce, D.A. (1986). "Enhancing the Performance of Large Diameter Piles by Grouting." Ground Engineering 19 (4) 7 pp and 19 (5) 7 pp.

COMPACTION GROUTING EFFECTIVENESS, A146, LOS ANGELES METRO RAIL

By Charles W. Daugherty,[1] Anthony F. Stirbys[2] and James P. Gould,[3] Honorary Member ASCE

Abstract: In 1988 and 1989 tunneling for the Los Angeles subway traversed a curve from Hill Street to 7th Street beneath an array of buildings. The designers intended compaction grouting to keep settlements of the building to an acceptable minimum during mining of the twin transit tunnels. To evaluate effectiveness of the compaction grouting, a test section was performed in a segment overlying the inbound tunnel on Hill Street before passing into the curve. After assessing the results, a second test section was carried out which included alterations to the tunnel mining procedures. This paper reports conclusions drawn from these test sections on the success of compaction grouting and mining methods in limiting settlements.

INTRODUCTION

Metro Rail Contract A 146 comprised excavation and lining of 640 m (2133 route feet) of twin tunnels, 6.7 m (22.3 ft) in excavated diameter. The tunnels follow a 300 m (1000 ft) radius curve from 5th/Hill Station to Seventh/Flower Station in the heart of downtown Los Angeles. Alignment of the tunnels with adjacent streets and buildings is shown in Fig. 1. Tunnels pass beneath twelve buildings ranging from a 3-story below-grade parking garage to several 12 to 14-story office towers. Depth of soil above the tunnel crown and below building

---

[1]Senior Professional Associate, Parsons Brinckerhoff, Quade and Douglas, 1 South Station, Boston, MA 02110.

[2]Senior Engineering Geologist, De Leuw Cather and Co., 523 West 6th St., Los Angeles, CA 90014.

[3]Consultant, Mueser Rutledge Consulting Engineers, 708 Third Avenue, New York, NY 10017.

footings is in the range of 7.5 m (25 ft) for the AR tunnel to 10.5 m (35 ft) for AL tunnel. Particular attention was focused on the reinforced concrete California Jewelry Mart beneath which the AL tunnel was to be driven first. The buildings were to remain fully occupied during construction. Specifications limited settlement from tunneling to 3 mm (1/8 inch).

Designers chose compaction grouting as the means of protecting the buildings. To evaluate its effectiveness and determine appropriate injection procedures, a test section 52 feet long was set up in April 1988 on Hill Street before reaching the curve. This was followed in October 1988 by a second test section which was used to assess not only compaction grout but also changes in mining procedures. Based on these results, protective measures and tunneling methods were planned in detail and successfully executed.

Tunneling Procedures

The contractor's excavator consisted of a Mitsubishi digger shield with backhoe bucket, one breasting table at the face's upper quarter point, seven half-moon jacks on the upper perimeter and two breasting jacks at springline. The initial tunnel liner consisted of four, 89° curved, pre-cast concrete segments, 0.23 m (9 in) thick, 1.2 m (4 ft) wide, expanded in the gap at the crown by hydraulic jacks. Screw jack spacers were installed and the crown gap filled with concrete. HDPE waterproof membrane was spread over the initial liner and then final lining was placed of cast-in-place concrete, 0.3 m (12 in) minimum thickness. Basic dimensions of shield and initial liner relating to ground loss are as follows:

Shield skin outside diameter =	6.62 m	(22.08 ft)
Overcutter thickness =	9.5 mm	(3/8 in)
Excavated cross-section area =	34.6 m^2	(385 sf)
Potential ground loss caused by:		
Overcutter annulus (3/8 in) =	0.2 m^3/m	(2 cf/ft)
Tail loss annulus (1.0 in) =	0.6 m^3/m	(6 cf/ft)
Volume recovered by expansion of liner, for each 25 mm (1 in) expansion =	0.1 m^3/m	(0.9 cf/ft)

GEOLOGIC SETTING

Overburden in A146 consists of Holocene alluvium in the buried valley of a stream tributary to an ancestor of the Los Angeles River. The sediments are typical of stream deposits in a semi-arid climate, varying from clays through silt and sand to large sizes. Groundwater

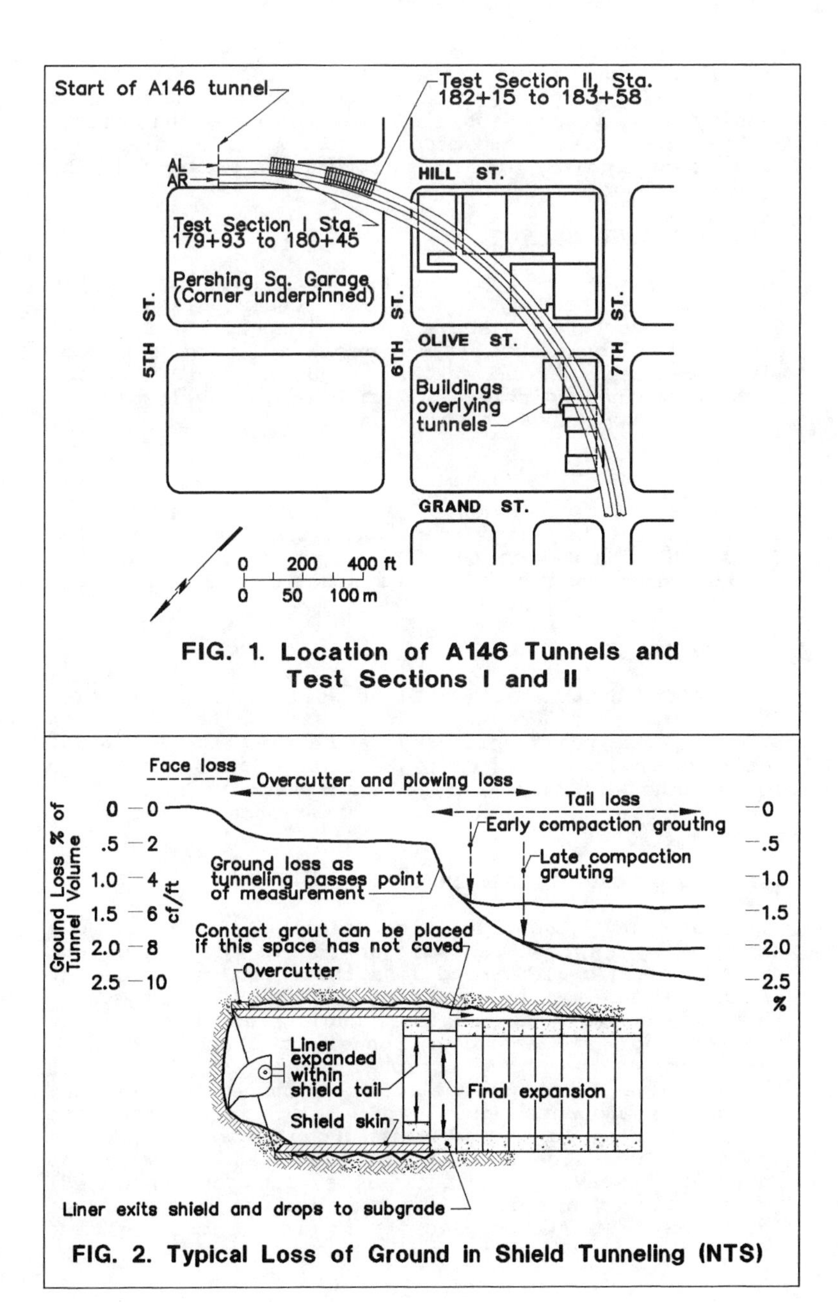

FIG. 1. Location of A146 Tunnels and Test Sections I and II

FIG. 2. Typical Loss of Ground in Shield Tunneling (NTS)

was 36 m (120 ft) deep, far below the tunnel, flowing on top of Miocene Fernando siltstone. The alluvium is exceptionally compact, probably near 100 percent relative density, having been densified by desiccation and seismic shakedown. All the deposit is clearly dilative. Constituent strata, identified roughly in order of depth, are as follows:

Stratum Y1: SW and GW;

Very dense, well graded sand and gravel with little silt and clay binder, lightly iron cemented, generally forming overburden above the tunnel. Standard penetration resistance (N) averaged 67 blows per foot with occasional refusal. It contains isolated Y2 lenses, a finer-grained, dirtier sand with some gravel which is of lesser importance in this tunneling.

Stratum Y3: SP and GP;

Dense to very dense, essentially cohesionless, uniform, fine to medium sand with some gravel and few fines. N values averaged 54 blows per foot. Y3 appears in the tunnel crown in certain reaches and is responsible for runs at the heading.

Stratum Y4: ML and CL;

Stiff to hard silt or clayey silt with interlensed fine sand. N values averaged 50. This alternates with Stratum Y3 in the tunnel crown and upper face. Typical shear strengths probably range from 6 to 8 ksf. The contrast between cohesionless Y3 and cohesive Y4 strongly influenced tunnel stability and performance of compaction grouting.

ELEMENTS OF GROUND LOSS IN TUNNELING

For this tunneling in relatively stable, dry alluvium by an open-face digger shield there are four factors which cause ground loss that lead to settlement of the surface or structures overlying (Hansmire 1985). These are commonly expressed as cubic feet of volume per running foot of length of tunnel or translated as a percentage of the cross-sectional volume (385 ft^2 or 34.6 m^2) for one foot advance of the tunnel excavation. Typical components based on job experiences are illustrated on Fig. 2 wherein ground loss is plotted against a cross section of the tunnel shield with the face being excavated and the initial liner being placed. Ground loss is exemplified by the absolute settlement of an extensometer anchor placed just above the tunnel crown. In Fig. 2 this time-settlement plot is translated to ground loss, apportioned as follows for stable ground:

	Percent Excavated Volume	Cubic ft per foot of Tunnel
1. Face loss due to overexcavation or runs in the heading.	0.1 to 0.2	0.4 to 0.8
2. Annulus opened by passage of the shield skin overcutter band which extends beyond the skin at the cutting edge.	0.5	2.0
3. Overexcavation beyond the skin created by plowing and yawing of the shield which varies from a smooth cut.	0.4	1.5
4. Tail loss created by incomplete expansion of the liner which fails to fill the annulus left when the tail of the shield moves forward.	0.5 to 1.0	2 to 4
Total	1.5 to 2.0%	6 to 8 cf/ft

In A146 the chief variable affecting ground loss was the effort employed to expand the segmented concrete liners to fill the annulus left by advance of the shield's tail. Compaction grouting is a means of offsetting one or another of these elements of ground loss. In less stable ground compaction grout usually is injected just at the time the heading reaches the pre-placed grout pipe. In A146 it was intended to be placed just after the final expansion of the pre-cast initial liner because the alluvium had been stable in the face.

Ground Loss Translated to Settlement

In dilative ground the area of the settlement trough at one diameter of cover is about 80% of the uncompensated tunnel ground loss and maximum settlement is numerically about 0.2% of that area (Hansmire 1985). Area of the settlement trough at two diameters of cover is about 60% of uncompensated tunnel ground loss and maximum settlement is 0.15% of that area. Thus if 6 cf/ft of uncompensated ground loss occurred in the tunneling, the settlement at one diameter of cover would equal 6 x .8 x .002 x 144 = 1.4 inches.

Specifications limited target settlement to 3 mm (1/8 inch) at the level of building footings. It appeared that compaction grouting might have to compensate for at least four cf/ft of ground loss, that is, one half of the potential ground loss. Certainly, tunneling procedures would have to be most rigorously controlled.

TEST SECTION I

This consisted of a 16 m (52 ft) length of AL tunnel in Hill Street which included five multiple-position extensometers and 22 compaction grout holes arrayed in 4 rows along the tunnel, roughly in a 2.7 by 2.7 m (9 by 9 ft) pattern. The opening at the tip of the grout hole was pre-formed to start grout bulbs about 3 m (10 ft) radially from the tunnel perimeter. Two tunnel cross-sections showing geology and position of grout pipes, grout bulbs and extensometer anchors are plotted in Fig. 3. Overburden above crown was 12 m (40 ft) and the clean sand of Y3 was everywhere in the crown.

Tunneling proceeded on the two, 8-hour shifts with a grouting crew on standby during the graveyard. Tunnel advance rate was typically 3 to 4 m (11 ft) in a shift which was still within the "learning curve". Advance increased to 6 m (20 ft) per shift on exiting the test section. Liner rings were assembled with a 18 cm (7 in) expansion in the crown inside the shield tail. Once the tail moved past an assembled ring, that ring was further expanded with a 55-ton jack in the crown gap to an average of about 28 cm (11-1/4 in). Full expansion of the crown gap to make liner segments completely occupy the excavated perimeter would have been 47 cm (18-1/2 in).

Grout take was irregular in compaction grout holes and 10 of the 22 refused grout. Certain holes which initially refused at pressures exceeding 600 psi were extended to within 5 feet of the liner. This was based on the condition that many grout bulbs were positioned in the hard Y4 silt which did not loosen but rather arched in response to tunneling. A distinctive feature was that grout injection took place typically three liner segments behind the shield tail, that is 3.6 m (12 ft) behind.

The compaction grout was specified to contain silty sand aggregate with 20 to 30 percent passing No. 200 sieve and a slump at a nominal 1-1/2 inches with 2 inches as a maximum. The mix actually placed included one sack of cement per yard and an aggregate slightly finer than specified to enhance pumpability. During an inspection, grouting consultant Ed Graf concluded that the mix was

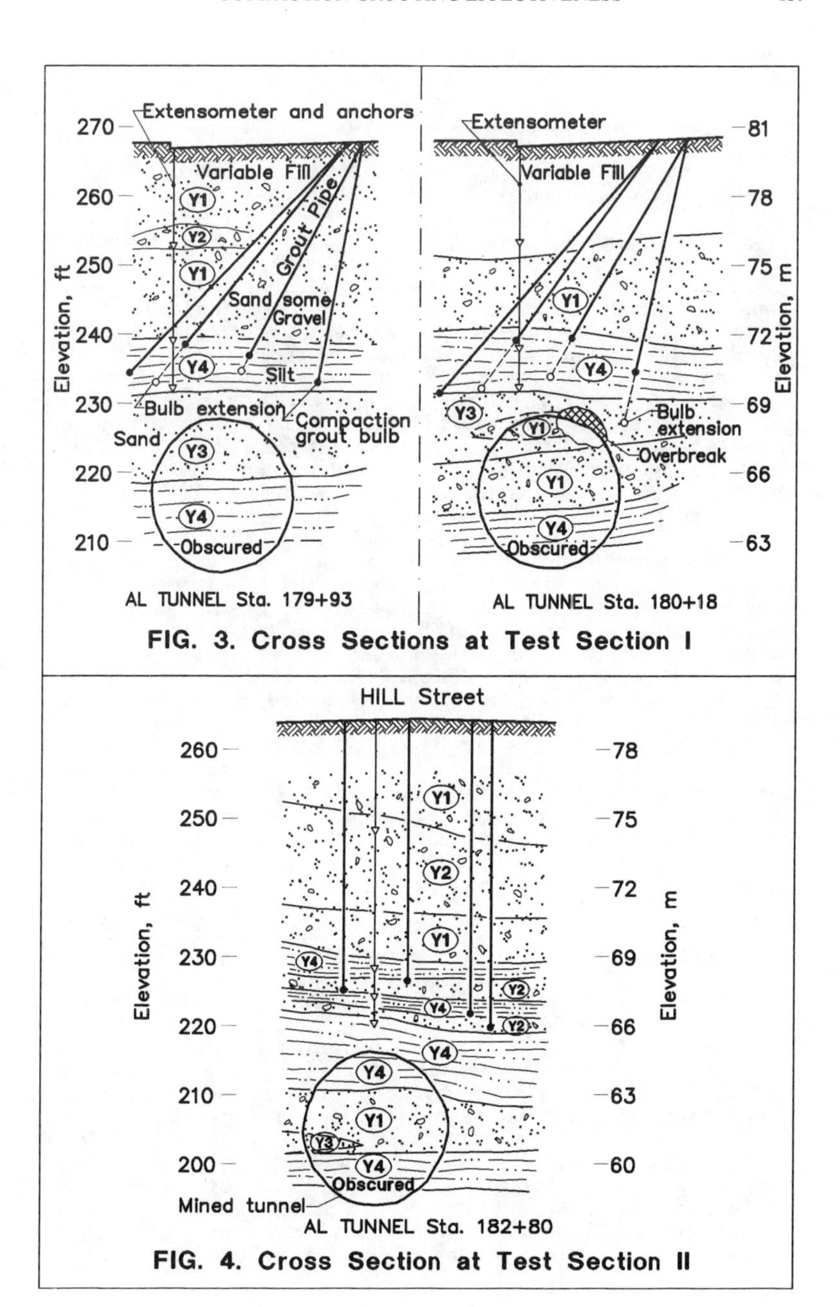

FIG. 3. Cross Sections at Test Section I

FIG. 4. Cross Section at Test Section II

too plastic and recommended mixing the supplied aggregate one to one with clean concrete sand.

Results of Test Section I

Observed results and interpreted values of ground loss are given in Table 1, modified from PDCD (1988).

This includes measured surface settlement and anchor settlements, liner expansion gap closure and the balance of ground loss volumes and compaction grout take volumes. Tunneling conditions were fair but with cohesionless Y3 sands in the crown throughout the test section, runs were experienced in the upper heading which were back-filled with sand bags. Although grout take was significant, averaging 6.9 cf/ft of tunnel, at least a portion was used in compacting the backfill placed in overbreak which totalled about 594 cf. Grout take ranged from 2 to 80 cf per hole and was markedly influenced by the soil in which the bulb was formed. The hard silt of Y4 would have had to dilate and tended to refuse grout.

The components of ground loss and the compensating volumes can be brought to reasonable agreement, as demonstrated in Table 1. Test Section I results indicate extraordinarily good ground control with the aid of the compaction grout which compensated for at least one-half of ground loss at the tunnel. Nevertheless, the 3 mm (1/8 in) target settlement was exceeded by a factor of three. As a result there occurred an interval of discussion and debate in the design team and between Authority and contractor. An intermediate length of tunnel was mined beyond Test Section I with results shown on Table 1. Various improvements in mining procedures were invoked and some decrease in shallow settlement was recorded. Specifications were modified to change the target limitation on settlement from 3 mm (1/8 in) to 25 mm (1 in).

TEST SECTION II

After these experiences, a second test section was established 43 m (143 ft) long within Hill Street, as shown on Fig. 1. This included a series of 5 subsections each one utilizing somewhat different details in the mining procedure. A typical cross section with compaction grouting pattern is shown in Fig. 4.

Attention was focused on Subsection D where 21 compaction grout holes were placed, most of these with bulbs started in hard silt of Stratum Y4. Grout mix and pressures were similar to Test Section I and 8 of the 21 holes refused grout. Average grout take was only 1.3

TABLE No. 1. Comparison of Results, Test Sections I and II

Observed or Estimated Factor		Test Section I 179+93 to 180+45	Intermediate 180+45 to 182+15	Test Section II 182+15 to 183+58
Observed Settlement above Centerline of Tunnel	Surface settlement	0.46" max 0.40" average	0.36" max 0.32" average	0.16" max 0.125" average
	Anchor 15' deep	0.51" max 0.42" average	0.41" max 0.35" average	0.17" max 0.13" average
	Anchor 5' above tunnel	2.4" max 1.48" average	1.2" max 0.91 average	0.52" max 0.39" average
Expansion of Liner at Crown Gap (% of Possible)		11-1/4" (61%)	12" (65%)	14-1/2" (78%)
Runs of ground		3, total 594 cf	none	none
Ground Loss Components: Cubic feet per tunnel foot	Face loss by runs not backfilled	3*	0.5*	0.4*
	Overcutter annulus, plowing & yawing	3*	1.5*	1.5*
	Incomplete liner expansion	6.5	5.8	3.6
	Total ground loss →	12.5	7.8	5.5
Factors Compensating Ground Loss: Cubic feet per tunnel foot	Volume of surface settlement	3 to 3.5	2.5 to 3	1 to 1.5
	Contact grout	0	0	1.2
	Compaction grout	6.9	2.2	1.3
	Dilation of overburden	2*	2*	0.5*
	Total compensa-tion →	12.4	7.2	5.0

* = Estimated values based on soil conditions and requirement for balance between ground loss and compensating factors.
(1" = 25mm; 1 cf/ft = 0.1 cubic meter / meter

cf/ft. Information on settlement and volume balances is presented in Table 1. In Subsection D crown stability was good, the hard Y4 silt arching above the shield. Because of this, 1.2 cf/ft of contact grouting could be injected even though only 1.3 cf/ft of compaction grout was placed. The most significant change in mining procedures was an additional effort at liner expansion with two 55-ton jacks expanding the crown gap, supplemented by jacks pushing liner outward at springline and at 1:30 and 10:30 clock positions. Average expansion was about 37 cm (14-1/2 in) leaving only a 10 cm (4 in) shortfall from full expansion, equal to ground loss of 3.6 cf/ft. These additional measures reduced volume in the settlement trough to between 1 and 1-1/2 cf/ft with an average settlement equal to target values. The mining was then judged to reach a satisfactory stage and excavation commenced beneath the block of buildings south of 6th Street.

COMPLETION OF THE TUNNELING

Final tunneling on the curve beneath the buildings was carried out with similar liner expansion and contact grouting, with compaction grouting from 720 holes placed as planned by the designer. Holes were predrilled on 3m (10 ft) centers and grout take was erratic as in the test sections. In the initial portion of the tunneling with Y4 silt in the crown, those procedures developed in Test Section II worked as expected. The tunneling passed beneath the first block of buildings without any significant settlement. However, as tunneling neared Olive Street, Y3 cohesionless sand again appeared in the crown and settlement at the parking lot increased to 25 mm (1 in). Concern arose that additional controls were necessary. The contractor proposed sodium silicate pattern grouting from the tunnel (Robeson and Wardwell 1991). That work was accepted as a change order and was carried out to provide a stable canopy at the crown. Excavation was then continued through to 7th Street with settlement limited to only a few millimeters as in the initial portion of the curve.

CONCLUSIONS

Los Angeles Metro Rail A146 provided an exceptional example of the selection of proper controls to tunnel beneath buildings with minimum settlement. This included not only use of compaction grouting, but later, chemical grouting to stabilize SP sands and rigorous controls of tunneling methods. Specific lessons are as follows:

- Where compaction grouting bulbs were expanded in granular soil which had been loosened by tunneling, they could be built up under high pressures. However, grout bulbs often could not be started in hard silt and clay. When the hard silt appeared in the crown there was sufficient arching to allow contact grouting to be injected outside the liner.

- When the face is stable and/or breasting is effective, compaction grouting is best timed to be injected just after the liner receives its final expansion outside the tail of the shield. In running ground, compaction grouting should be injected much earlier, as the heading passes beneath the grout pipe. Liner expansion must be as rapid and highly powered as practical.

- The volume of compaction grout that could be placed in the cohesionless sand, several cubic feet per running foot of a single tunnel, offset a substantial portion of the ground loss volume. However, improvement in tunneling procedures could achieve a roughly comparable reduction of ground loss. One inch of added circumferential expansion of the liner represents almost one cubic foot per foot of reduction in ground loss. In controlling tunneling all elements must be assessed: face control, liner expansion and injecting grout in various forms. No single measure is a panacea to control settlement but all must be considered and applied as ground conditions dictate.

APPENDIX: REFERENCES

Hansmire, W.H. (1985), "Supplemental Report, Influence of Tunneling and Requirements for Building Protection, Section A140, SCRTD - Metro Rail Project," July 26, Los Angeles.

Parsons, De Leuw Cather, Dillingham (PDCD) (1988), "Report on the Compaction Grouting Test Section, Station 179+95 to 180+45 AL, Contract A146, SCRTD-Metro Rail Project," August 25, Los Angeles.

Parsons, De Leuw Cather, Dillingham (PDCD) (1989), "Report on the Test Section II, Station 182+15 to 183+58 AL, Contract A146, SCRTD-Metro Rail Project," October 27, Los Angeles.

Robeson, M.J. and Wardwell, S.R. (1991), "Chemical Grouting to Control Ground Losses and Settlement on Los Angeles Metro Rail, Contract A146," UTRC Proceedings.

Evaluation of Cone Penetrometer Grout Backfilling

by Scott M. Mackiewicz[1] and John R. Rohde[2]

ABSTRACT

Cone penetrometer testing (CPT) is quickly becoming an accepted method of *in situ* environmental site investigation. Besides providing continuous measurement of soil parameters, recent advances have allowed *in situ* measurement of soil resistivity and contaminant fluorescence. Cone systems have also been developed for environmental sampling of vadose zone soil gases and pore water. As with any penetration test, removal of the device after testing creates a potential vertical pathway for contaminant migration. Several grouting methods for utilization with CPT have been developed. This study focuses on development of improved equipment utilizing a grout hose and outlet located above the cone, and a new methodology that allows a measurement of grouting effectiveness during cone equipment removal. The concept behind the method is to use a pressure transducer, already commonly used for pore water pressure measurement on the cone, to maintain a constant grouting overpressure during extraction of equipment. Based on grout shear viscosity and applied overpressure, this methodology provides a potentially valuable verification measurement that the grout is adequately sealing the void. The suitability of the grouts, capable of being pumped through a narrow grout hose, for backfilling material was determined by their physical properties of density, settlement, and viscosity. An initial laboratory test of the technique was used to forensically determine the efficiency of the method. The result of this study indicates that the method of injecting grout at a constant overpressure while varying the extraction rate through materials of different permeability could assure uniform grout placement. The experimental values agree with Raffle and Greenwood's equations for modeling spherical divergent two-phase grout flow, and these equations allow predictions of grout volumes required for closure.

[1] Project Engineer, GeoSystems Engineering, 7856 Barton, Lenexa, KS 66214
[2] Associate Professor, Dept. of Civil Eng., Univ. of Nebraska, Lincoln, NE 68588

INTRODUCTION

Cone penetrometer testing (CPT) is becoming an accepted method of *in situ* soil investigation. Besides providing continuous measurement of soil parameters (Robertson and Campanella, 1986), recent advances have allowed *in situ* measurement of soil resistivity and contaminant fluorescence. Cone systems have also been developed for environmental sampling of vadose zone soil gases and pore water. Upon test completion, the rods and cone are extracted leaving behind a vertical void in the soil. When used in environmental applications, this vertical void is of concern because it creates a potential pathway for vertical contamination spread. A process for sealing the hole immediately after completion of a CPT soil investigation would greatly increase the potential for use of CPT in environmental investigations. This paper presents a laboratory evaluation of a method and the processes and materials used for electronic cone penetrometer hole closure.

GROUT PENETRATION IN SOILS

The rate of flow of the grout into soils is dependent on various characteristics including the viscosity and shear strength of the grout, the soil permeability, and the placement pressure (Bowen, 1981). Flow of Newtonian grouts into soils has been modeled using a spherically divergent two-phase flow equation (Raffle and Greenwood,1961):

$$t = \frac{na^2}{kh}\left[\frac{\Psi}{3}\left(\frac{R^3}{a^3}-1\right)-\frac{\Psi-1}{2}\left(\frac{R^2}{a^2}-1\right)\right] \tag{1}$$

where t = time (T); n = porosity (L^3/L^3); k = soil permeability (L/T); h = pressure head (L); a = radius of source (L); R = influence radius of grout (L); Ψ = grout/water viscosity ratio. The relationship between the radius of grout penetration, the source radius, injection time, and grout viscosity, is shown in Figure 1 (Raffle and Greenwood, 1961). Once a radius is known at a specific time; the volume flow rate needed to reach the radius can be determined, Equation 2, from the same soil and grout characteristics (Raffle and Greenwood, 1961).

$$h = \frac{Q}{4\pi k}\left[\Psi\left(\frac{1}{a}-\frac{1}{R}\right)+\frac{1}{R}\right] \tag{2}$$

This model has limitations, for example in layered soils the flow is less divergent and consequently flow rates are overestimated. This limitation, however, does not have much effect on the Newtonian grouts used today. With the aid of these expressions, a rate of Newtonian grout injection and dissipation can be evaluated when all grout and soil characteristics are known.

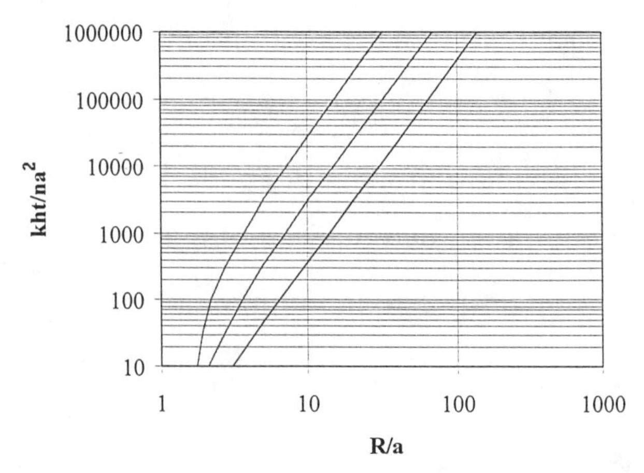

Figure 1. Radius of penetration for various grout/water viscosity ratios (after Raffle and Greenwood, 1961)

For Non-Newtonian grouts, limitations arise in the radius of influence equation due to their structured form. The influence radius is reduced from the drag corresponding to the grout's shear strength. The drag from soil shear stress will ultimately become the same magnitude as the applied placement pressure and grout will no longer maintain viscous flow. An estimation of the drag coefficient for different soils can be determined by (Raffle and Greenwood, 1961):

$$\alpha = \sqrt{\frac{8\eta_{pl} k}{n\gamma_w}} \tag{3}$$

where α = drag coefficient (L); η_w = water viscosity at 20° C (MT/L^2); γ_w = unit weight of water at 20° C (M/L^3); n = porosity of the soil (L^3/L^3).

Using the relationship of the drag coefficient, a limiting radius of influence can be determined if the yield shear stress of the grout, placement pressure and source radius are known by (Raffle and Greenwood, 1961)

$$R_L = \frac{\gamma_w h\alpha}{20\tau_f} + \alpha \tag{4}$$

where τ_f = yielding shear stress (M/L^2) and R_L = limiting radius (L).

The understanding of Newtonian and Non-Newtonian flow through soil provides a better picture of how injection procedures work. This information provides the background for the development of a grout injection system for CPT.

EXPERIMENTAL METHODS

Experimental grouts consisted of Non-Newtonian (i.e., particulate) grouts consisting of bentonite, cement(Type I), and combination bentonite and cement(Type I). A total of nine mixes were designed with varying water to cement ratios, bentonite percentages, and cement percentages. A tabulation of all mixes and their constituents are shown in Table 1. It should be noted that in the general use of a combination of cement and bentonite permits a higher bentonite percentage in the grout while still being pumpable. The density of the grout material is tabulated for reference of its regulatory compliance according to mud weight.

Table 1. Experimental Grouts

	Percentage of Constituents by Weight				
Sample Number	Cement (c)	Bentonite (b)	Water (w)	Ratios	Specific Gravity
Cement (Type I) Grouts					
#1E	33.3	---	67.7	w/c=2	1.264
#2E	25	---	75	w/c=3	1.210
#3E	20	---	80	w/c=4	1.156
Bentonite (Pure Gold Powder) Grouts					
#4E	---	5	95	---	1.034
#5E	---	6	94	---	1.036
#6E	---	7	93	---	1.037
Bentonite (Pure Gold Powder)/Cement (Type I) Grouts					
#7E	8.75	3.5	87.75	b/c=0.40	1.098
#8E	14	3.5	82.5	b/c=0.25	1.140
#9E	14	7	79	b/c=0.50	1.150

A high speed blade mixer was used to thoroughly mix each of the experimental grouts. Initial pumping tests were performed using an air over grout cell that was attached to a 0.635 cm I.D. hose. This small diameter hose was chosen on the basis that it was small enough to be fitted within standard cone push rods. The maximum pressure used during pump test was below 551 kPa. These initial pump tests were conducted to determine if the mixes were pumpable through a cone penetrometer hole backfilling system. All grouts listed in Table 1 were pumpable, to some extent, through the small diameter tube. Other experimental methods used for determination of a feasible grout include the settlement, viscosity, and soil penetration tests.

Settlement Tests

The settlement test involved filling graduated cylinders with 500 ml of grout and recording the level of the particulate solution with time. A measurable decrease in particulate solution level indicates settlement and may indicate that the grout would be problematic in field applications and potentially not provide a continuous seal near the top of the test hole depth due to excess water. In some cases, if sedimentation occurs and the excess water disappears through dissipation and dehydration, prediction of the ultimate uniformity of the placed grout is questionable. Thus, one of the criteria for the grout mixture design is that only a minimal amount of grout sedimentation is allowed to insure an adequate and reliable seal.

Viscosity Testing

The two main characteristics that define Non-Newtonian fluids are apparent plastic viscosity and yielding shear stress. These two properties were evaluated using a Haake Rotovisco concentric cylinder rotary viscometer. The viscometer allows the bob to be rotated at different fixed speeds, 3.6 to 583.2 r.p.m., which permits investigation over a wide shear range. This viscometer is designed to shear a fluid in an annulus between two concentric cylinders, one which is rotating and the other is held stationary (Skelland, 1967). The fluid is placed within the non-rotating cylinder called the cup. The rotating cylinder called the bob which is suspended by a torsion wire and attached to the viscometer is placed within the cup. The torque used to rotate the bob is measured by a scale for various rates of shear. The scale value can be translated into shear stress and viscosity by knowing various dimensions and constants of the bob and cup cylinders. The dimensions are measured directly from the cylinders and the constants are calculated during calibration tests using an oil of known viscosity. The cup is filled with either 40 or 70 ml of sample depending on which bob type is used. The cup is then inserted into the drive mechanism creating an air trap with the hollow recess on the underside of the bob. This air trap prevents the sample from coming in contact with the bottom of the rotating bob, thus minimizing possible end effects.

The test is performed by reading the speed factor of the rotating cup (U) and the scale (S) which measures torque on the bob. These values along with the given shear factor for the bob (B) and a calibration constant (K) are used to calculate the shear rate (D) on the rotor surface, , and the shear stress(τ) using

$$\textit{Shear Rate(D) in sec}^{-1} = \frac{\textit{Rate of Shear Factor for Bob(B)}}{\textit{Speed Factor(U)}} \tag{5}$$

$$\textit{Shear Stress}(\tau) = \frac{\textit{Calibration Constant(K)} * B * \textit{Scale Reading(S)}}{100} \tag{6}$$

Once the shear stress (τ) and shear rate (D) were calculated, a shear rate versus shear stress curve was derived. This curve allows interpolation of the yield shear stress value from the x-axis intercept and calculation the apparent plastic viscosity from its slope. These two relationships were found for time increments of 0, 0.5, 1, 2, 3, 4 hours after mixing to better understand the grout changes over time. The changes in yield shear stress and plastic viscosity affect the radius of grout influence to be modeled.

Soil Permeation Tests

The soil permeation test was performed to find the influence radius of grout penetration into soils under applied pressures and under its own gravimetric head. The permeation test was performed by uniformly compacting a cylindrical barrel with a clean medium sand of a known permeability (k = 1.5E-02 cm/s) around an injection PVC pipe with a 5.08 cm outside diameter. The injection PVC pipe was buried 1 meter in the barrel. The injection PVC pipe was fitted with a 16.4 m length of 6.35 mm I.D. tube to simulate field cone penetrometer backfilling conditions for shallow holes. The tubing was attached to a pressure gage to measure the backfilling pressure, see test setup schematic in Figure 2. Grout was placed within the pressure vessel that was constructed of schedule 40 PVC pipe and valves. The cell has a top valve for filling of the vessel and a bottom valve to control flow while under pressure. A sight tube was mounted on the vessel to allow measurement of the amount of grout left in the vessel at a specific time. A regulated compressed air tank was connected just below the top valve in order to pump at the regulated air pressures using an air over grout method. With the top and bottom valves closed, the applied pressure was raised until equalization. The bottom valve was opened allowing for grout flow through the 15.24 m length of tubing and out of the exit port. The grout was placed at an exit pressure of 17.23 kPa to produce a steady intrusion of grout with a small influence radius.

Upon test completion, the grout was left to solidify. After solidification, the uninfluenced soil was excavated to retrieve a grout column. The grout column diameter was measured along its length allowing for comparison of experimental and theoretical values. The extraction time which varied throughout the test was recorded allowing for calculations of radius of influence for different depths of the sample. The influence radius and limiting radius of the grout was predicted using the measured placement pressure, viscosity, yielding shear stress, source radius, soil permeability, and time. A correlation between the theoretical and experimental values was determined to evaluate Raffle and Greenwood's equations.

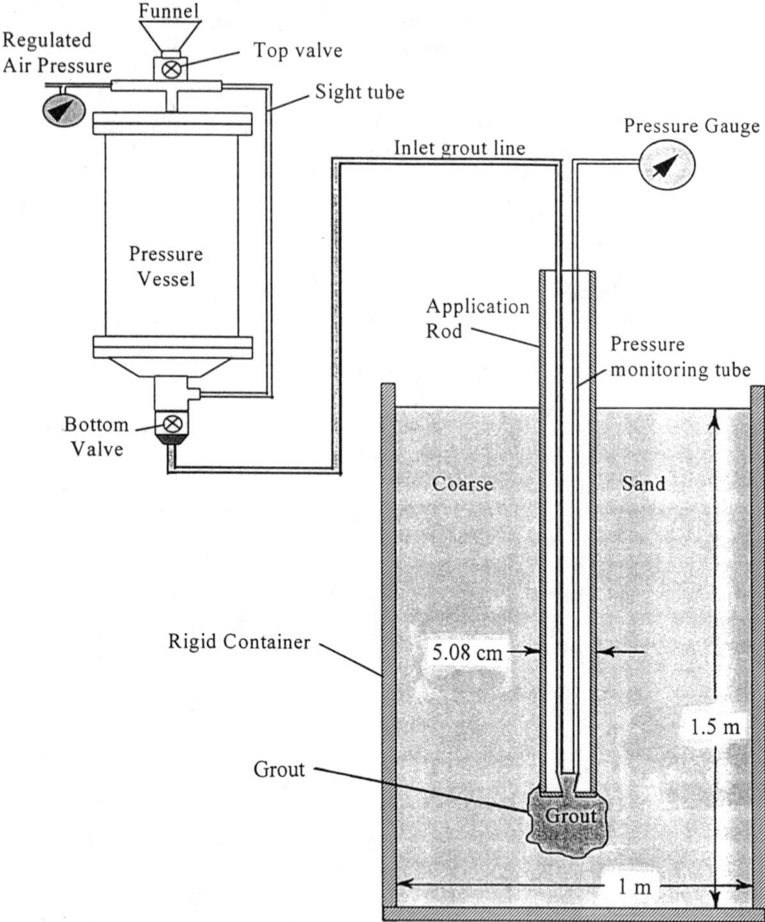

Figure 2 Schematic of soil permeation system.

EXPERIMENTAL RESULTS

Settlement Test Results

The results of the settlement test showed that the cement particulate grouts segregated almost immediately, as shown in Table 2. On the other hand, the clay and cement/clay suspension had very little or no segregation over a 24 hour period. From these test results, the cement grouts(samples #1E-#3E) were not used in the remainder of the experiments. The tests show that for a 24 hour period the settlement decreases with increasing bentonite/cement ratio(b/c) from 50 ml for a b/c ratio = 0.25 in Sample #8E to a settlement of 20 ml for a bentonite/cement ratio = 0.50 in Sample #9E. Settlement tests results show that by increasing the bentonite percentage the cement particles stay suspended making the grout more stable.

Table 2. Settlement Test Results, Amount of Settlement in (ml)

Sample	1 Minute	15 Minute	30 Minute	24 Hours
#1E	25 ml	170 ml	180 ml	190 ml
#2E	50	235	240	240
#3E	70	300	300	300
#4E	0	0	0	0
#5E	0	0	0	0
#6E	0	0	0	0
#7E	0	0	5	25
#8E	0	5	20	50
#9E	0	5	10	20

Viscosity Test Results

The results of the viscosity test are useful in determining the flow rates and grout dissipation into soil pores. The shear rate versus shear stress curve were evaluated for the apparent plastic viscosity and yielding shear stress for each sample at six different time periods; 0, 0.5, 1, 2, 3, 4 hours. The test data indicates that the plastic viscosity of all samples increases with time, see Figure 3. The test data also revealed, that the yielding shear stress increases over time due the grouts thixotropic properties, see Figure 4. It was concluded from test results that sample #9E was optimum for the soil permeation tests due to its large yielding shear stress(>150 dynes/cm^2) that in turn will reduce the influence radius and reduce material needs.

Soil Permeation Test Results

After excavation, the measurements of the grout column were recorded and tabulated, a summary is provided in Table 3. The measured results of the experimental test were compared to the theoretical and limiting radius values calculated from the Raffle and Greenwood's equations. The theoretical values were

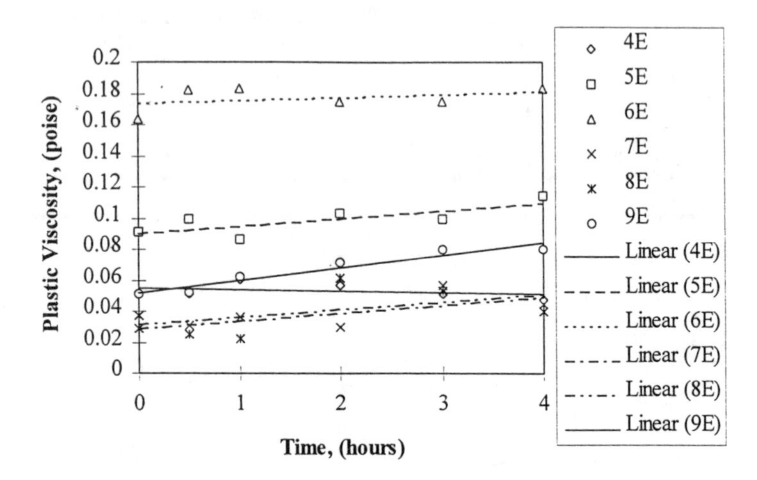

Figure 4. Plastic Viscosity values of tested grouts with relation to time.

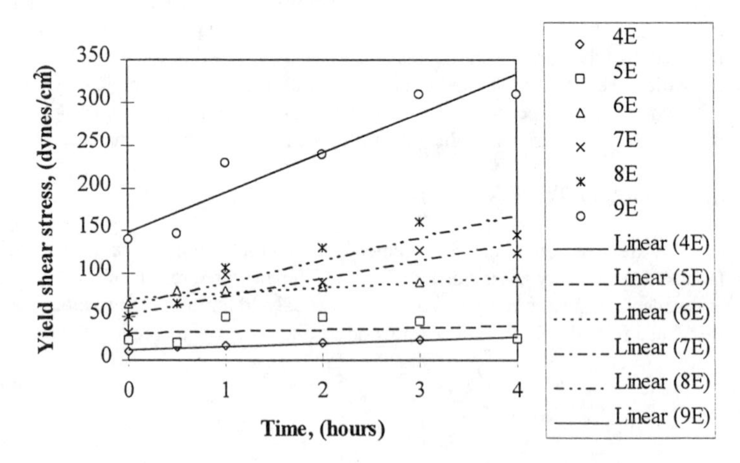

Figure 4. Yield shearing stress of grouts with relation to time.

calculated using the changing grout properties including its plastic viscosity and yielding shear stress over time from the viscosity test results. The calculations for the influence and limiting radius according to Equations 1 & 4 are compiled in Table 3 along with the difference between the experimental and limiting radius results being presented as percentage error.

Table 3. Soil permeation test tests results

Depth of Specimen (cm)	Experimental Radius (cm)	Limiting Radius (cm) (EQ. 4)	Theoretical Radius (cm) (EQ. 1)	Error (%)
0	----	2.92	2.77	----
15.2	2.82	3.01	11.96	6.31
30.5	2.65	3.11	15.26	14.79
48.3	2.76	3.22	18.13	14.29
63.5	3.25	3.32	20.36	2.11
83.8	3.58	3.46	22.78	3.47

CONCLUSIONS

This work has presented a method for closing the vertical void left behind by electronic cone penetrometer testing. The experimental testing performed provided an understanding of design parameters for performing and evaluating small diameter tube backfilling systems. The testing incorporated a specific grouting material, mixed with a high speed blade mixer, and a grout placement monitoring method for the cone penetrometer.

During experimental testing, a grouting material was chosen for its ability to seal the hole, ease of mixing, and pumpability through a narrow tube, for which the size is restricted by cone rod diameter. Out of the nine particulate grouts tested, a grout containing 7% bentonite and 14% portland cement by weight was used during the soil permeation test. The experimental grout's low rate of settlement, along with its fluidity and high yield shear stress without admixtures made it an optimum mixture for hole closure under the established criteria.

A grouting verification method was incorporated by injecting the grout with a constant volumetric rate and regulating the extraction rate after the specific placement pressure had been reached. Varying the rate of cone extraction while monitoring the placement pressure should allow different permeability material to be grouted effectively. The experimental values for a medium sand material support this procedure and allow modeling and estimation of needed resources.

The results of the soil permeation test, performed on a medium sand material, indicate that Raffle and Greenwood's equations can be used to model the grout radius of influence. Using these equations, estimations of grout volume and time for hole closure may be determined along with cost projection and efficiency calculations for this innovative *in situ* grouting procedure.

APPENDIX I. REFERENCES

Bowen, Robert. Grouting in Engineering Practice. 2 ed. New York: John Wiley & Sons, 1981.

Raffle, J.F. and Greenwood, D. A. "The Relation Between the Rheological Characteristics of Grouts and their Capacity to Permeate Soil." Proceedings of the 5th International Conference on Soil Mechanics and Foundation Engineering. 1961.

Robertson P.K. and Campanella R.G. Guidelines for Use and Interpretation of the Electronic Cone Penetration Test. 3rd ed. Gaithersburg, Maryland: Hogentogler & Company, Inc., November 1986.

Skelland, A. H. P. Non-Newtonian Flow and Heat Transfer. New York: John Wiley & Sons, Inc., 1967.

APPENDIX II. NOTATION

The following symbols are used in this paper:

a = radius of source (L);
h = pressure head (L);
k = soil permeability (L/T);
L = length of capillary (L);
n = porosity (L^3/L^3);
P = applied pressure (M/L^2);
Q = volume of flow (L^3/T);
r = radius of capillary (L);
R = influence radius of grout (L);
t = time (T);
α = drag coefficient (L);
τ_f = shear stress at which material starts flowing (M/L^2)
ε = thickness of liquid film through which plug flows (L);
Φ = liquid film fluidity (L/MT);
Ψ = grout/water viscosity ratio;
η_{pl} = plastic viscosity (MT/L^2);
η_w = water viscosity at 20° C (MT/L^2);
γ_w = unit weight of water at 20° C (M/L^3);

SUBJECT INDEX

Page number refers to first page of paper

AUTHOR INDEX

Page number refers to first page of paper